Praise for *The Dyslexic Advantage*

"This brave book helped lay the groundwork for the neurodiversity revolution—for seeing conditions like dyslexia, autism, and ADHD as alternate styles of learning and being that convey benefits as well as challenges. This updated edition is even more informative, provocative, and wise."
 —Steve Silberman, author of the *New York Times* bestseller *NeuroTribes*

"*The Dyslexic Advantage* is a true celebration of diversity of thought. In it, the Eides use science and real-world examples to demonstrate that—when it comes to human information processing—different does not necessarily mean better or worse. In fact, any understanding of the shadow of difficulties associated with dyslexia is incomplete without an exploration of the light that creates those shadows. And as it turns out, the brain designs associated with dyslexia can be quite advantageous when tasked with the right problems. I highly recommend it for anyone interested in how different people learn and problem-solve, regardless of where you might personally fall on these dimensions of thinking!"
 —Chantel Prat, PhD, author of *The Neuroscience of You*

"*The Dyslexic Advantage* shows you how to celebrate rather than suffer from your dyslexia."
 —*The Times* (London)

"*The Dyslexic Advantage* helps dyslexics and their families recognize and nurture the benefits of a dyslexic brain."
 —*Wired*

"[*The Dyslexic Advantage*] underscores how important it is for leaders to learn about experiences that are not our own and to lean in to differences in how people think and work. . . . An important reminder that too often we don't focus on people with disabilities—let alone people with invisible disabilities—when we talk about diversity, equity, and inclusion, and that needs to change."
 —Rachel Thomas, cofounder and CEO of Lean In, for CNBC.com

"I can't recommend this book highly enough."
 —Jessica Lahey, parenting and education writer and author of the *New York Times* bestseller *The Gift of Failure*

"*The Dyslexic Advantage* is a paradigm-shifting book that captures the remarkable advantages that come with this different style of thinking. This book should be the first people reach for when they want to learn about what it really means to be dyslexic."
—Vince Flynn, *New York Times* bestselling novelist

"A fantastic read for anyone who wants to learn about dyslexia. Brock and Fernette Eide have a knack for explaining complex ideas and scientific work in a simple manner *and* offering great insights. This is probably the most helpful material ever published on dyslexia. . . . A classic."
—Manuel Casanova, MD, professor of biomedical sciences at the University of South Carolina School of Medicine, Greenville

"This provocative book explores the misunderstood side of dyslexia that is characterized by skill and talent. The authors focus on what dyslexic people do well. This is a must-read for parents, educators, and people with dyslexia."
—Gordon F. Sherman, PhD, former president of the International Dyslexia Association

"A compelling call to action."
—*Scientific American Mind*

"Gripping, powerful, and insightful—and for dyslexics, immensely validating."
—Nancy Ratey, EdM, senior disability analyst and author of *The Disorganized Mind*

"Groundbreaking theory for the positive potential of dyslexia . . . Good advice and encouraging analysis."
—*Kirkus Reviews*

"Here's a practical analysis of a difficult, frustrating disorder that unveils what goes remarkably right as well as what goes wrong."
—*Publishers Weekly*

THE DYSLEXIC ADVANTAGE

(REVISED AND UPDATED)

Unlocking the Hidden Potential of the Dyslexic Brain

Brock Eide, MD, MA, and
Fernette Eide, MD

PLUME

An imprint of Penguin Random House LLC
penguinrandomhouse.com

Copyright © 2023 by Brock L. Eide and Fernette F. Eide
Penguin Random House supports copyright. Copyright fuels creativity, encourages diverse voices, promotes free speech, and creates a vibrant culture. Thank you for buying an authorized edition of this book and for complying with copyright laws by not reproducing, scanning, or distributing any part of it in any form without permission. You are supporting writers and allowing Penguin Random House to continue to publish books for every reader.

PLUME and P colophon are registered trademarks of Penguin Random House LLC.

LIBRARY OF CONGRESS CATALOGING-IN-PUBLICATION DATA

Names: Eide, Brock, author. | Eide, Fernette, author.
Title: The dyslexic advantage: unlocking the hidden potential of the dyslexic brain / Brock L. Eide, MD, MA, and Fernette F. Eide, MD.
Description: Revised and updated. | New York: Plume, 2023. | Revised edition of: The dyslexic advantage / Brock L. Eide and Fernette F. Eide. New York: Hudson Street Press, c2011. | Includes bibliographical references and index.
Identifiers: LCCN 2022040024 (print) | LCCN 2022040025 (ebook) | ISBN 9780593472231 (paperback) | ISBN 9780593472248 (ebook)
Subjects: LCSH: Dyslexia—Psychological aspects.
Classification: LCC RC394.W6 E33 2023 (print) | LCC RC394.W6 (ebook) | DDC 616.85/53—dc23/eng/20220831
LC record available at https://lccn.loc.gov/2022040024
LC ebook record available at https://lccn.loc.gov/2022040025

Printed in the United States of America
3rd Printing

Book design by Tiffany Estreicher

While the authors have made every effort to provide accurate telephone numbers, Internet addresses, and other contact information at the time of publication, neither the publisher nor the authors assume any responsibility for errors or for changes that occur after publication. Further, the publisher does not have any control over and does not assume any responsibility for author or third-party websites or their content.

Neither the publisher nor the authors are engaged in rendering professional advice or services to the individual reader. The ideas, procedures, and suggestions contained in this book are not intended as a substitute for consulting with your physician. All matters regarding your health require medical supervision. Neither the authors nor the publisher shall be liable or responsible for any loss or damage allegedly arising from any information or suggestion in this book.

To Krister, for teaching us the importance
of focusing on strengths.

To Karina, in loving memory.

And to dyslexic people everywhere.

Contents

Preface to the New Edition   xi
Introduction   xv

Part 1. A Matter of Perspective
1. A New View of Dyslexia   3
2. Dyslexia from Two Perspectives   11
3. Modeling Dyslexia   22

Part 2. The Scientific Dimensions of Dyslexia
4. Sources of Dyslexia's Dual Nature   33
5. More Conscious, Less Automatic   49
6. Dyslexia-Associated Challenges   60

Part 3. M-Strengths: Material Reasoning
7. The "M" Strengths in MIND   75
8. The Advantages of M-Strengths   82
9. Trade-offs with M-Strengths   94
10. Growing Up with M-Strengths   102
11. Key Points About M-Strengths   109
12. What's New with M-Strengths   113

Part 4. I-Strengths: Interconnected Reasoning
13. The "I" Strengths in MIND   127
14. The Advantages of I-Strengths   134

15. Trade-offs with I-Strengths   146
16. Brains Without Borders   154
17. Key Points About I-Strengths   159
18. What's New with I-Strengths   164

Part 5. N-Strengths: Narrative Reasoning

19. The "N" Strengths in MIND   181
20. The Advantages of N-Strengths   186
21. Trade-offs with N-Strengths   202
22. A Story for Every Occasion   207
23. Key Points About N-Strengths   211
24. What's New with N-Strengths   216

Part 6. D-Strengths: Dynamic Reasoning

25. The "D" Strengths in MIND   239
26. The Advantages of D-Strengths   246
27. Trade-offs with D-Strengths   252
28. D-Strengths in Business   259
29. Key Points About D-Strengths   266
30. What's New with D-Strengths   273

Part 7. Bringing the Dyslexic Advantage to Life

31. Exploratory Strengths of the Dyslexic Mind   297
32. Educating Dyslexic MINDs   315
33. Dyslexic MINDs at Work   339
34. Modeling the Dyslexic Self   361

Epilogue   377

Acknowledgments   383

Appendix: Mind Strengths Survey Results   387

Notes   397

Index   413

THE DYSLEXIC ADVANTAGE

(REVISED AND UPDATED)

Preface to the New Edition

It is now almost twelve years since the first edition of *The Dyslexic Advantage* was published. When we wrote the first edition, we hoped to make a few simple points:

- That the true significance of dyslexia will never be understood if we look only at dyslexic reading and spelling challenges, and fail to look at the kinds of things dyslexic minds do really well;

- That both dyslexia-associated strengths and challenges appear to be produced by the same brain organizational and cognitive features, like two sides of the same neurobiological coin; and,

- That the key to helping individuals with dyslexia find success and personal fulfillment in their learning, work, and life is not just to fix their areas of weakness, but to help them to understand, develop, and enjoy their strengths.

This view of dyslexia differed radically from the conventional view at that time, so we were uncertain how our book would be received. Fortunately, we have been thrilled at the reception these ideas have received not only from dyslexic individuals and their families, but also from the educational, professional, and academic research communities. From the former we have received thousands of messages of thanks and appreciation, many describing the book's life-changing impact; from the latter we have received many expressions of interest, including invitations to speak or to collaborate on new projects. Our book has also led to the formation of the Dyslexic Advantage nonprofit organization (dyslexicadvantage.org), which for the last decade has provided information and encouragement to dyslexic individuals and their families, and to the creation of Neurolearning SPC (neurolearning.com), which provides high-quality dyslexia screening for individuals ages seven and older.

Almost twelve years on, our book continues to sell as well as it did its first year, both in the United States and abroad. It has also played an important role in encouraging a growing international movement focused on identifying, building, and engaging the strengths of dyslexic individuals.

Why, then, is an update needed? Certainly not to retract anything we said earlier. Quite the opposite. The case for dyslexia-associated strengths is now stronger and clearer than ever. That's why it's time for an update. There is simply so much more we can now say in support of a dyslexic advantage.

New research, much of it performed using recently developed techniques, has further clarified the nature of the strengths associated with dyslexia and connected them in surprising ways. This work yields new insights on how dyslexic minds function, why they are so common in the human community, and why so many dyslexic individuals show the kinds of strengths we describe. Even more important, it provides us with a clearer perspective on what dyslexic minds are actually organized and optimized to do.

This new edition includes:

- Significantly updated chapters on the neurobiology underlying dyslexia and dyslexic strengths;

- "What's New" chapters for each of the MIND strengths, highlighting important recent research that we have conducted ourselves, as well as relevant studies that have been released in the past twelve years;

- Updates within the previously existing chapters where needed;

- A new framework to create a fuller, more balanced, and more useful model of dyslexia;

- An explanation for why we now believe that dyslexic minds play a deeply beneficial role in our human community;

- Entirely revised chapters on school, work, and self-concept that focus on the implications of a strengths-centered model of dyslexia both for dyslexic individuals and for their educators, employers, and coworkers;

- Approximately twenty new interviews with an outstanding and diverse group of dyslexic individuals, who share generously from their own lives and experiences and highlight the nature of the MIND strengths.

For our dyslexic readers using the print edition, our publishers at Plume have also graciously made this version more accessible by printing it in a larger and more dyslexia-friendly font.

So join us in taking a new and updated look at the dyslexic advantage. Whether you are dyslexic yourself; the parent, spouse, partner, teacher, or employer of a dyslexic person; or simply someone who is interested in learning more about the amazing ways that human minds work, we believe you will find this journey as inspiring, fascinating, and richly rewarding as we have.

Introduction

In 2004, a top business school in England sent out a press release with the headline: "Entrepreneurs Five Times More Likely to Suffer from Dyslexia." The release went on to ask what made some of England's most well-known and successful entrepreneurs like Richard Branson, Alan Sugar, and Anita Roddick so special. The answer, according to Julie Logan, who headed the research described in the press release, was that each of these highly successful entrepreneurs also "suffered from dyslexia," a condition she and her colleagues found to predispose quite strongly to entrepreneurial success.

In light of the tremendous success enjoyed by these entrepreneurs, it seems rather odd to describe them as "suffering from dyslexia." Yet as almost anyone with dyslexia can tell you, being dyslexic truly can involve a great deal of suffering: like the shame of constantly failing at skills others master with ease; the ridicule of peers and classmates; or exclusion from classes, schools, or careers one would otherwise pursue. Yet it is equally clear when we look closely at individuals with dyslexia—when we see how they think, and what they can do, and the often remarkable persons they become—that in many respects "suffering from dyslexia" is suffering of a most unusual kind.

This book isn't about dyslexia. It's about *the kinds of individuals who are diagnosed with dyslexia*: the types of minds they have, the ways they process information, and the things they do especially well. It's not a book about something these individuals *have*. It's about who they *are.*

Most books on dyslexia focus on problems with reading and spelling. While these problems are extremely important, they are not the only—nor even the most important—things that distinguish dyslexic from nondyslexic minds. In this book, we will focus primarily on these additional differences, which we believe hold the key to understanding what dyslexia and dyslexic minds are really all about.

As experts in neuroscience and learning disabilities, we've had the privilege of working with thousands of dyslexic individuals and their families. In the process we've found that these individuals tend to share a broad range of important cognitive features. Some of these features are learning or processing challenges—like difficulties with reading, spelling, rote math, working memory, or visual and auditory function. However, others are important strengths, abilities, and talents. We call these abilities the *dyslexic advantage.* While these features differ somewhat from person to person, they also form recognizable patterns—just like the different musical works of Mozart are distinguishable from one another, yet recognizably the work of the same composer.

Traditionally, attempts to understand dyslexia have focused almost entirely on problems with reading, spelling, and other academic skills. As a result, little attention has been paid to the things individuals with dyslexia do especially well—particularly once they've become adults. In our opinion, this is a grave mistake. Trying to understand what dyslexia is all about while overlooking the talents that mature individuals with dyslexia characteristically display is like trying to understand caterpillars while ignoring the fact that they grow up to become butterflies.

As we'll show you in this book, the brains of dyslexic individuals aren't defective; they're simply different. Their differences in wiring often lead to special strengths in processing certain kinds of information, and these strengths typically more than make up for the better-known dyslexic challenges. By learning how to recognize, nurture, and properly use these strengths, individuals with dyslexia can be helped to achieve greater success and personal fulfillment.

There are two big differences between the traditional view of dyslexia and the one we'll present in this book. First, we don't see the academic challenges associated with dyslexia as the result of a disorder or a disease. Instead, we see these challenges as arising from a different pattern of brain organization—one whose chief aim is to predispose dyslexic individuals to the development of valuable skills that aid in understanding and adapting to a complex and changing world. When dyslexia is viewed from this perspective, we can see that the strengths and challenges that accompany it are like two sides of the same neurobiological coin. In this book, we'll identify these advantages, describe how they can be used, and explain why we believe that they—rather than challenges with reading and spelling—should be seen as dyslexia's true core features.

Second, unlike most books on dyslexia, this book won't focus on making individuals with dyslexia into better readers. Instead, it will focus on helping them to become better *at being dyslexic*. Reading instruction can change certain brain features in ways that will help dyslexic persons become better readers. This is a good thing, and something we strongly support. Yet it's also important to recognize that reading instruction doesn't change *all* the things that make dyslexic brains different from nondyslexic ones—and this is also a good thing, because dyslexic brains aren't *supposed* to be like everyone else's. Dyslexic brains have their own kinds of strengths and abilities, and these advantages should be recognized and enjoyed. Our goal is to help individuals with dyslexia recognize these many wonderful advantages so they can enjoy the full range of benefits that can come from having a dyslexic mind. The first step in achieving this goal is to help them think more broadly about what it really means to be dyslexic by expanding the concept of dyslexia so that it no longer means only challenges but also includes important talents.

The best way to broaden our perspective on dyslexia is to look not only at things that individuals with dyslexia find challenging, but also at the kinds of things they often do especially well. One obvious way to do this is by studying people who've excelled at being dyslexic. Most instructional books or videos on topics like playing sports, cooking, or speaking foreign languages have one thing in common: they feature expert practitioners sharing and modeling tips and strategies they've personally found useful. Since this is a book about how to excel at being dyslexic, we'll share lots of stories, tips, and pointers from dyslexic individuals who've enjoyed success in their own lives. While not every individual with dyslexia will succeed in precisely the same ways as these talented individuals, anyone with a dyslexic processing style can benefit from their insights and from studying the strategies they've used.

In the early chapters of this book, we'll describe how dyslexic brains differ from nondyslexic ones. Then we'll examine each of the four dyslexia-associated strength patterns we've found to be common in individuals with dyslexia. We've called these patterns the *MIND strengths* to make them easy to remember. These four MIND strengths are *m*aterial reasoning, *i*nterconnected reasoning, *n*arrative reasoning, and *d*ynamic reasoning. These strength patterns are not meant to be rigid or watertight categories; instead, they are helpful ways of thinking about and understanding dyslexic talents. In fact, during the years since the first edition of this book was published, it has become apparent how closely connected each of these different talents is.

While none of the MIND strengths is exclusive to individuals with dyslexia, each is linked to a set of particular cognitive and structural brain features common in individuals with dyslexia. As you read these chapters, please remember that although most individuals with dyslexia share many features, each individual is also unique. Dyslexic processing isn't caused by a single gene, so different individuals with dyslexia will show different patterns of strengths and challenges. Very few will show all the MIND strengths, but almost all will show some.

We hope that this book will be a source of both useful information and encouragement for all those who have not yet learned the many wonderful advantages that can come from being dyslexic.

Part 1

A Matter of Perspective

1
A New View of Dyslexia

Throughout his school career, Doug struggled with reading and writing. He flunked out of community college twice before he finally gained the skills he needed to earn his college degree. Today he's the president of a highly successful software firm that he founded two decades ago.

When Lindsey was young, all her teachers called her slow. Although she worked desperately to learn to read and write, she was one of the last in her school class to master these skills. Although she barely made it through elementary school, Lindsey not only graduated from college but also earned the top prize in her school's highly competitive honors program. She's now a clinical psychologist.

Pete's elementary school teachers told his parents he was borderline mentally retarded and emotionally disturbed. They also told them that they couldn't teach him to read or write. However, with intensive one-on-one instruction, Pete learned to read and write well enough not only to attend college but also to go on to law school. Pete eventually used his legal training to represent another individual with dyslexia before the Supreme Court, winning her case 9–0 and radically redefining the rights of students with special educational needs.

Doug, Lindsey, and Pete are all dyslexic, and they're also all exceptionally good at what they do. As we'll show you in this book, these facts are neither contradictory nor coincidental. Instead, Doug, Lindsey, and Pete—and millions of individuals with dyslexia just like them—are good at what they do not *in spite of* their dyslexic processing differences, but *because* of them.

This claim usually provokes surprise and a flurry of questions: "Good because of their dyslexia? Isn't dyslexia a learning disorder? How could a learning disorder make people good at anything?"

A learning disorder couldn't—if it was *only* a learning disorder. But that's just our point, and it's the key message of this book. Dyslexia, or the *dyslexic processing style*, isn't just a barrier to learning how to read and spell; it's a reflection of an entirely different pattern of brain organization and information processing—one that predisposes a person to important abilities along with the well-known challenges. This dual nature is what's so amazing—and confusing—about dyslexia. It's also why individuals with dyslexia can look so different depending on the perspective from which we view them.

Look first at individuals with dyslexia when they're reading or spelling or performing certain other language or learning tasks. From this perspective they appear to have a learning disorder; and with respect to these tasks, they clearly do. Now look at these same individuals when they're doing *almost anything else*—particularly the kinds of tasks they excel at and enjoy. From this new perspective they not only cease to look disabled but often appear remarkably skilled or even specially advantaged.

This apparent advantage isn't just a trick of perception—as if their strengths seemed large only in contrast with their weaknesses. There's actually a growing body of evidence supporting the existence of a *dyslexic advantage.* As we'll discuss throughout this book, studies have shown that the percentage of dyslexic professionals in fields such as engineering, art, and entrepreneurship is over twice the percentage of dyslexic individuals in the general population. Individuals with dyslexia are among the most eminent and creative persons, like singer-songwriters Carly Simon and John Lennon; *Shark Tank* entrepreneurs and venture capitalists Daymond John, Barbara Corcoran, and Kevin O'Leary; Nobel Prize–winning scientists Carol Greider and John Goodenough; Academy Award–winning filmmakers Spike Lee, Steven Spielberg, and Octavia Spencer; inventors Yoky Matsuoka and Dean Kamen; architects Johnpaul Jones, Richard Rogers, and Michel Rojkind; and legendary product designers Chuck Harrison and Jony Ive.

Importantly, the link between dyslexic processing and special abilities isn't visible just among superachievers. You can prove this for yourself by performing a simple experiment. Next time you run across an unusually good designer, landscaper, mechanic, electrician, carpenter, plumber, radiologist, surgeon, orthodontist, small-business owner, computer software or graphics designer, photographer, artist, boat captain, airplane pilot, or skilled member of any of the dozens of dyslexia-rich fields we'll discuss in this book, ask if that person or anyone in his or her immediate family is dyslexic or had trouble learning to read, write, or spell. We'll bet you dollars for dimes that person will say yes—the connection is just that strong.

Now, would these connections be possible if dyslexia was *only* a learning disorder? The answer, clearly, is no. So there must be two sides to dyslexia. While dyslexic processing clearly creates challenges with certain academic skills, these challenges are only one piece of a much larger picture. As we'll describe throughout this book, dyslexic processing also predisposes individuals to important strengths in many mental functions, including:

- three-dimensional spatial reasoning and mechanical ability
- the ability to perceive connections and relationships like analogies, metaphors, paradoxes, similarities, differences, and missing pieces
- the ability to remember past personal experiences in vivid detail and to use fragments of these memories to perform all sorts of cognitive tasks
- the ability to perceive subtle patterns in complex and constantly shifting systems or data sets, and to mentally simulate and predict the outcome of complex processes over time

While the precise nature and extent of these abilities vary from person to person, there are enough similarities between these strengths to form a recognizably related set. Together these talents can be referred to as *dyslexia-related abilities* or a *dyslexic advantage*. Remarkably, these abilities appear to arise from some of the same variations in brain structure, function, and development that give rise to dyslexic challenges with literacy, language, and learning.

In this book we'll argue for a radical revision of the concept of dyslexia: a sort of Copernican revolution that places abilities rather than disabilities at the center of what it means to be dyslexic. This shift in perspective should change not only our thinking about dyslexia but also the ways we educate, employ, and teach individuals with dyslexia to think and feel about themselves, their abilities, and their futures.

Please understand that we're not in any way trying to downplay the hardships that individuals with dyslexia often experience or to minimize their need for early and intensive learning interventions. We are simply trying to expand the view of dyslexic processing so that it encompasses both the challenges that individuals with dyslexia often face and the abilities they commonly demonstrate. This broadened perspective can be illustrated using the following analogy.

A Tool Discovered

Imagine you live on a remote island and you've never had contact with the people or products of the outside world. One morning as you walk along the beach you spy a shiny cylindrical tube half buried in the sand. You pick it up, clean it, and carefully examine it. With growing excitement, you realize it's a product of human design; but what it is or what it's for, you can't immediately decide.

As you inspect the tube, you find that it's roughly as long as your arm and as heavy as a fist-sized stone. It's also gently tapered so that one circular end is nearly twice as wide as the other. You turn the tube to inspect its large end, and as you bring it toward your eye, you see that a light is shining through it. You bring it still closer, then peer cautiously through this large end. After a moment's adjustment you begin to see a familiar yet marvelously transformed image: It's a lovely, delicate miniature of the beach stretched out in front of you. With awe and amazement, you realize what you've discovered: a remarkable device for making things look small!

Well, yes and no. . . .

Like a telescope, the concept of dyslexia is a human invention, and like a telescope it can either expand and clarify our view of individuals who struggle to read and spell or, used the wrong way around, it can cause our view of these individuals to shrink. Unfortunately, this diminishing effect is what has happened with dyslexia for far too long.

How the Narrow View Became the Primary View

Surprisingly, given that we now know that dyslexic processing is present in as many as one in five individuals, the first clear description of an individual with dyslexia appeared in the medical literature just over a century ago. In 1896, British ophthalmologist W. Pringle Morgan described a fourteen-year-old boy named Percy. Despite receiving seven years of "the greatest efforts . . . to teach him to read," Percy could read and spell at only the most basic level, even though his schoolmasters believed he was "the smartest lad in the school."

It was through this case that the concept of dyslexia was first developed: the idea that there exists a distinct group of individuals who—though clearly intelligent—learn and process the information encoded in printed language very differently from their nondyslexic peers. Historically, the processing features most commonly associated with dyslexia are difficulties with reading and spelling, though, as we'll see later, other challenges with language and learning are also common in individuals with dyslexia.

While the development of this concept of dyslexia has been useful, we believe its true worth has never been fully realized because, as was true with the telescope in our example, one critical question has been overlooked: *How should this telescope be used?* Should it be used as a tool to narrow our view solely to literacy, language, and learning difficulties? Or should it be turned around so we can see *all* the learning and processing features of this amazing group of individuals: not just in literacy and language but across the whole range of their activities—strengths as well as challenges—and throughout their entire life span?

Since this question has largely gone unasked, challenges with literacy, language, and other aspects of learning have remained the almost exclusive focus of dyslexia research and education. As a result, dyslexia has come to be seen as essentially synonymous with those challenges. This perspective is reflected in current definitions of dyslexia. In the United States, the most widely used definition was developed by the National Institute of Child Health and Human Development (NICHD) and subsequently adopted by the International Dyslexia Association (IDA). It reads:

> Dyslexia is a specific learning disability that is neurological in origin. It is characterized by difficulties with accurate and/or fluent word recognition and by poor spelling and decoding abilities. These difficulties typically result from a deficit in the phonological component of language that is often unexpected in relation to other cognitive abilities and the provision of effective classroom instruction. Secondary consequences may include problems in reading comprehension and reduced reading experience that can impede growth of vocabulary and background knowledge.

In the terms of our telescope analogy, this is clearly a shrinking perspective. It narrows our view of dyslexic processing to the challenges experienced by individuals with dyslexia, and it does nothing to expand our view of their skills or capacities. From this perspective, dyslexia is merely:

- a *learning disability*
- characterized by *difficulties*
- resulting from *deficits*
- that produce *secondary consequences*
- and additional *impediments*

No wonder people have such a negative view of dyslexia!

But is there really any reason to believe that this definition tells us everything we need to know about individuals with dyslexia? In a word: no. Habit alone has led us to assume that the first use we discovered for this telescope is the only one, and habit alone keeps us from discovering other—and better—uses. Because we first recognized dyslexia as a *learning disorder* rather than a *learning* or *processing style*, we've paid little attention to whether dyslexic processing might also create talents and abilities. However, as we'll show you in the next chapter, the talents and benefits that are associated with dyslexic processing can be easily observed once we recognize that dyslexia can be viewed from two different perspectives.

2

Dyslexia from Two Perspectives

To demonstrate the enormous difference that results when dyslexia is viewed from these two different perspectives, we'd like to introduce you to a family we've been privileged to meet through our work.

We first met Kristen when we spoke to a parents' group about the challenges we often see in very bright children. After our presentation, Kristen introduced herself and told us about her son. Christopher was in third grade, and he'd recently shown very broad gaps in his performance on the different subtests used in IQ testing. While he'd scored well on tests measuring verbal and spatial reasoning, his performance was weaker on tests measuring processing speed and working memory (or mental desk space, which we'll discuss later). Kristen wanted to know if we could tell her anything about children who show such a pattern.

We told Kristen that this is a pattern we often see in bright young children, whom we've affectionately dubbed our "young engineers." Although many of these young people show strong interest in verbal subjects such as history, mythology, fantasy literature, role-playing games (including game creation), reading or being read to, and even storytelling or creation of imaginary worlds, they typically show their keenest interest in spatial or mechanical activities like building, designing, art, inventing, electronics, computing, and science. We told her that many of these children struggle with handwriting, written expression, spelling, and initially with reading (especially with oral reading fluency). Often they are persistently slow readers, and a smaller number show challenges with oral language expression, like word retrieval, or difficulty putting their thoughts into words. Many also have a clear family history of dyslexia or have many relatives who've excelled as adults in occupations requiring spatial, mechanical, or higher mathematical skills.

Kristen at first appeared surprised by our remarks—almost taken aback—and we wondered if we'd missed our mark. But after a short silence she smiled and said, "Let me tell you about my family."

One Dyslexic Family: The Narrow Perspective

Kristen's son, Christopher, showed his first dyslexia-related challenges quite early in development. Like many dyslexic children, he was slow to begin speaking (his first words came shortly after his second birthday) and slow to combine words into sentences. As a preschooler, he struggled to find words to express his thoughts, and his speech was often unclear. He often subtly mispronounced words and confused similar-sounding words like *polish* and *punish*. Despite the fact that he could identify the numbers 0 through 10 before his second birthday, Christopher couldn't learn the names of letters until he was almost five years old. In school, he was much slower learning to read and write than most of his classmates, and he also had great difficulty memorizing math facts, despite a strong grasp of number concepts and math reasoning.

Christopher received special testing and was referred to several learning specialists, who helped him with reading, handwriting, speech articulation, and word retrieval. His reading accuracy had improved by the time he reached fourth grade, but he still struggled with slow written work production, messy handwriting, and spelling.

Kristen, too, showed many signs of dyslexia early in her life. She was very slow in learning to read, and according to her parents she still struggled with basic phonetic decoding as late as fourth grade. Like Christopher, she made frequent word substitutions (like *peaches* for *pears*), struggled to get her thoughts down in writing, and spelled very poorly. Kristen also had a weak memory for auditory or verbal sequences—like phone numbers or word spellings—and she had difficulty mastering abstract verbal concepts that she couldn't easily picture.

Kristen recalled that during her early years in school she was often "incredibly bored" and "couldn't stand desk work." She found listening to lectures on abstract subjects especially difficult. During much of middle school and the early years of high school, she received low grades and came close to failing. She eventually realized that time was flying by and that "if she wanted to get anywhere in life" she'd have to go to college, so she buckled down and was able to raise her grades enough to get into a state university.

Kristen initially planned to major in sociology or psychology, which were the subjects she found most interesting. However, she soon realized, "If I had to read or write for my degree, I wouldn't make it through." Instead, she majored in interior design, and after earning her degree, she went to work for a large design firm.

Kristen's father, James, doesn't remember any unusual difficulties learning to read as an elementary school student in the 1930s. However, throughout his life he has exhibited a persistent discrepancy between his high intellectual ability and his difficulty learning from text, which is characteristic of individuals with partially compensated dyslexia. He has always been a slow reader, has never read for pleasure, and according to his family was able to succeed in high school and college largely because his childhood sweetheart—and later his wife of nearly seventy years—Barbara, helped him do the reading for his coursework. Until her recent death, she still helped him with his business-related reading.

Handwriting, spelling, and written expression also troubled James throughout his education, and they've remained tough for him to this day. Kristen fondly recalls the time when she made an imaginary diner from a cardboard box and asked her father to help her spell *restaurant* above its door. He thought for a moment, then said, "You should call it a *café* instead—that's a much nicer word." James also had difficulty remembering math facts like the times tables (a problem he's never entirely overcome); remembering math equations and certain rules and procedures; taking notes during lectures; grasping verbal (and especially abstract verbal) concepts; mastering a foreign language; switching his attention from one subject to another as required at school (despite good prolonged attention for his preferred activities); and feigning interest in almost everything they were trying to teach him in school, with the exception of his advanced science courses.

This was the *narrow* view of Christopher, Kristen, and James that emerged when we focused exclusively on their dyslexia-related challenges. Now let's see how they appeared when we broadened our focus to look instead at their strengths.

One Dyslexic Family: A Second Look

Christopher, when we first met him at age nine, already displayed many of the talents and abilities that are common among individuals with dyslexia. He showed very strong three-dimensional spatial abilities that have revealed themselves in several ways since early in life. For example, when Christopher was only three, his family was staying at a large hotel that had been formed by combining several older and very different structures into one enormous complex. After checking in at the front desk, the family walked for several minutes through a confusing labyrinth of passages to reach their room. Once there, they deposited their bags and walked out again in search of dinner. When they returned to the hotel several hours later, Christopher announced that he would lead the family back to their room. To his parents' astonishment, he did so without a single hesitation or mistake.

Christopher's spatial abilities also revealed themselves in his persistent love of building. Although he enjoyed using just about any kind of building material, LEGOs were a special favorite, and he often spent hours using them to build complex and unique designs in a room in his house devoted solely to that purpose. In addition, he displayed a passionate interest in science and how things work.

Christopher also showed many impressive verbal strengths, despite some persistent focal language challenges. He'd always had a great love of stories, and even before he could speak, he would listen with rapt attention to the reading of lengthy stories, like *The Velveteen Rabbit*. When his reading skills improved, he become a voracious reader, and he read for both entertainment and information. Despite his difficulties with verbal output, on his IQ testing, Christopher had strong verbal scores.

Kristen similarly appeared much different when we looked at her from this "broadened" perspective. Although her school career was marked by difficulty remembering many kinds of abstract verbal facts, Kristen's memory was remarkably good in other ways. By mentally "tapping out" numbers on an imaginary keyboard, Kristen could recall a long list of phone numbers, including those of many places where she'd worked or lived and friends' numbers stretching back into childhood.

Kristen, like Christopher, had a very strong spatial sense and could quickly and indelibly learn her way around new environments. She had a phenomenal visual memory for people and places from her past and could still easily "see" where she sat in all her school classes as a child, who sat around her, what many of them wore, and how the walls of her classrooms were decorated.

Many—perhaps most—of Kristen's memories had a strong contextual, personal, or "episodic" element, involving elements of past experience. Kristen experienced these memories like dramatic scenes playing out in her mind. They portrayed information about where she first encountered each fact, object, individual, fashion, song, or other remembered item, including whom she was with, what she saw or heard, and how she felt. Kristen had a similar sensory-immersive experience of sounds, colors, touch sensations, and emotions when she read or heard stories. As we'll see in later chapters, this type of vivid experiential (or *episodic*) memory is extremely common in individuals with dyslexia, and it is often accompanied by weaknesses in rote and abstract verbal memory.

As with many of the dyslexic students with whom we work,

Kristen found learning to be a very personal, almost intimate experience. During her years in school, this made her learning highly dependent upon her relationship with her teachers and her interest in her course materials.

While Kristen's memory skills were impressive, they did little to help her with her schoolwork—though, channeled properly, they clearly could have. Instead, her vivid recall of personal experiences often created a powerful inducement to daydream. As a result, it took a great deal of interesting "outside" stimulation from her teachers to hold her attention.

Eventually, however, these cognitive traits became the foundation for Kristen's highly successful career. After finishing college, Kristen went to work for a firm that designed and furnished office spaces. She quickly became one of the most productive design and sales representatives in this nationwide operation. Kristen credits much of her professional success to her spatial and personal memory skills, which allow her to imagine how interior spaces would look when changed in various ways. She also found that her restless energy, drive, and dislike of desk work—all of which made it so hard for her to sit passively in class every day—were ideally suited for a job that required visits to construction sites, suppliers' showrooms, and clients' offices, as well as frequent phone calls to troubleshoot issues.

Kristen's father, James, also discovered that many of the cognitive features that troubled him in school became keys to his success in the working world. While he showed few early signs of promise inside the classroom, outside of school he displayed a remarkably bright and precocious mind. At age six he built his first radio-controlled boat, which he designed to include a special compartment to carry his lunch. He spent much of his time "taking things apart" to see how they worked, and his interest in electronics was further piqued when an electrician visited his home to install a new stove and took the time to demonstrate his tools and techniques.

James also developed an interest in magnetism. When he was a student during World War II and dozens of sand buckets were brought to his school for fire safety, James demonstrated that the sand was full of iron by running a magnet through the buckets. Rather than receiving encouragement for his interest in experimental science, James was disciplined for "playing" in the sand.

In tenth grade James finally found a science teacher who could answer his insightful questions and who was able to lead him into new and deeper areas of interest. Chemistry and physics became special passions, and he reveled in the pleasure of having a teacher who saw him as especially promising. In response, James not only completed all the required reading but also labored through several more advanced books. Outside of school, James strengthened his knowledge of electronics by working for the electrician who had earlier befriended him. During the summer after his junior year of high school, James put this knowledge to work by building a commercial-grade AM radio station, which he sold to a local entrepreneur.

After earning his degree in physics at Reed College, James went to work for Battelle Memorial Institute in Richland, Washington. There he quickly distinguished himself as a talented and creative inventor. He received his first patent for an electron beam welder, which he followed with a steady stream of inventions that he created to solve problems for clients.

However, James's most famous invention had its origin not in a client's problem but in one of his own. James has always loved classical music, and he loved to play his favorite recordings again and again and again. Back when his music was stored on LP records, he was driven nearly crazy by the hisses, scratches, and skips that accumulated when he played his favorite records repeatedly. Seeking to eliminate the wear that came from repeated physical contact between a stylus and a grooved vinyl record, James imagined a system in which an optical reader would detect digital information embedded on a small plate with which it never made physical contact. Over the next several years, James invented the seven components that together became known as the compact disc system. The impact of this invention on data storage and retrieval—not just for music but for all types of information—has been profound. In fact, you'll find James T. Russell's compact disc system on many lists of the most important inventions of the twentieth century.

Dyslexia and Talent: An Essential Relationship

The lives of James, Kristen, and Christopher—though in some ways unique—display many of the features we commonly see in individuals with dyslexic processing styles. In fact, it was seeing these patterns again and again in the dyslexic families with whom we work that convinced us that certain strengths are as much a part of the dyslexic profile as challenges in reading and spelling.

Please notice that we're *not* saying merely that individuals with dyslexia can be talented *in spite of* their dyslexia—like James Earl Jones overcoming his stutter to become a great actor or Franklin D. Roosevelt overcoming polio to become president of the United States. Instead, we're claiming that certain talents are as much a part of dyslexic processing as the better-known challenges; or in other words, that the strengths and the challenges are simply two sides of the same neurobiological coin. We can explain what this connection is like using an example from the sport of baseball. Consider the following players:

Barry Bonds	Rafael Palmeiro
Hank Aaron	Reggie Jackson
Babe Ruth	Jim Thome
Willie Mays	Mike Schmidt
Ken Griffey Jr.	Manny Ramirez
Sammy Sosa	Mickey Mantle
Frank Robinson	Jimmie Foxx
Mark McGwire	Frank Thomas
Alex Rodriguez	Willie McCovey
Harmon Killebrew	

If you're at all familiar with baseball, you'll recognize at least a few of these names, and if you're a real fan, you'll recognize them all. These are some of the greatest stars who've ever played the game. Consequently, you may be surprised to learn that another thing they have in common is that they are all among the top one hundred all-time leaders in striking out while at bat!

This seems like a rather unwelcome distinction, since striking out is unquestionably a kind of mistake—like misspelling a word or misreading a sentence or writing illegibly. If you knew nothing more about these players other than that they'd struck out more than almost all the other batters in major-league history, you'd probably conclude that they were poor players. You might even conclude that they shared a kind of hitting disability, like "dysfunctional batting syndrome" or "contact deficit disorder."

However, one thing is certain: if you didn't also know that when we first prepared this list in 2011, these players were also the top nineteen home-run hitters of all time, you'd have a highly incomplete—and seriously misguided—impression of their value and ability as players. Although striking out is clearly undesirable in itself, when it is viewed in the context of the game as a whole, we discover that even the greatest players strike out a lot if they swing hard enough and often enough to hit a lot of home runs. Since avoiding strikeouts isn't as important in baseball as scoring runs—which these big-swinging home-run hitters did extremely well—the list of strikeout leaders turns out to be a list of some of baseball's greatest winners, not its losers. The strikeouts are simply a kind of trade-off that are accepted for the sake of scoring more runs.

This relationship between home runs and strikeouts is a lot like the connection between the strengths and challenges in dyslexia. The home runs that dyslexic brains have been structured to hit are *not* perfect reading and spelling, but skills in other kinds of complex processing that we'll discuss throughout this book. It's because dyslexic brains have been organized to make these home runs possible that they're also at higher risk for striking out when they try to decode or spell words.

Discovering these strengths—and how they can be used to help individuals with dyslexia find success in the classroom and the workplace—is what this book is all about. To fully understand these dyslexia-associated strengths and how they fit together with dyslexia-associated challenges, we will need to form an entirely new model of what dyslexia truly is. In the next chapter we will begin our discussion of what such a model should look like.

3
Modeling Dyslexia

We want to start this section by telling you about our friend Gerry Rittenberg. Gerry is very smart. He's got an incisive mind that quickly cuts to the essence of whatever we talk to him about, and he never misses the forest for the trees. Gerry has also amassed a remarkable track record over many years with many kinds of businesses.

Gerry's first job out of college was as a problem-solver and labor/management go-between at a large packaging factory. He did very well and was repeatedly promoted, but he felt confined within the corporate structure and eventually left to form his own business selling party goods. One of his clients in that business was so impressed that he invited Gerry to join him as a partner. A few years later he made Gerry his successor.

In the mid-1990s, Gerry took over as CEO of the party supplies wholesaling business Amscan. At the time it had annual revenues of $50 million. By the time Gerry retired just over twenty years later, he had grown Amscan to annual revenues of more than $3 billion. During those years he acquired over two dozen other companies, yet he made sure his business never became so complex that he could not summarize its operations—along with his plan and vision for its future—in a half page or less. Gerry actually kept such a summary on his desk at all times, just to remind himself to maintain his focus—but he didn't really need it. As he told us many times, "I can just *see* it—I can see in my mind how it all fits together."

Gerry's greatest business success started out as his biggest challenge. The retail firm that was Amscan's biggest client declared bankruptcy. Gerry knew that if the bankruptcy went through it would hurt—maybe even destroy—Amscan. Instead of focusing on the danger, Gerry stepped back and saw opportunity. He went to a private equity group with a bold proposal. If they would join him in buying his bankrupt retail client, he'd take over the client's operations himself. Gerry promised he would bring the company back to profitability within two years. The investors knew Gerry's work and liked his plan, so they said yes. But the plan didn't quite go as predicted. Instead of taking two years to become profitable, Gerry's new acquisition became profitable in only nine months. A decade later Gerry orchestrated the highly successful public listing of that company, Party City, on the New York Stock Exchange.

Clearly, Gerry is smart. However, there was a time in Gerry's life when there seemed to be little evidence of the brilliant strategic thinker he's become. All through primary and secondary school, Gerry struggled to pass his courses. Reading and writing were torture. When he took the SATs, his combined score from both parts fell below 800. At college he received work-study credits for managing the student social club, which got him out of a lot of classes. But when he did venture into the classroom, his problems with reading and writing continued.

For the many people who mistakenly think *smart* means

"someone who does well in school," individuals like Gerry are confusing. Yet Gerry has proved his intelligence in countless ways over a long span of time. What Gerry is *not* is someone who fits easily into simple ideas about what it means to be smart. That's because Gerry is dyslexic, and dyslexia is not simple.

One of the best pieces of evidence we can give you that dyslexia is not simple is that Gerry himself—a person who can seamlessly describe the workings of several dozen companies in a half-page summary—could never find a description of dyslexia that he felt had the same clarity. We met Gerry shortly after the first edition of this book came out, and one of the first things he asked us was the apparently simple question "What *is* dyslexia?" Gerry had been told for over forty years that he was dyslexic, and he had asked dozens of experts to define dyslexia, but he'd never received an answer that made sense of all the challenges and strengths that he believed went along with this complex thinking style.

Gerry found our description in *The Dyslexic Advantage* helpful. He even joined the board of our Dyslexic Advantage nonprofit and became one of its biggest supporters. But every so often he'd ask us again, "What *is* dyslexia?" Over time we came to realize that what Gerry was really asking was "What does it actually mean to be a person with dyslexia, and to have a dyslexic kind of mind?" And what he was looking for wasn't a book-length explanation or a list of strengths, but a single image or a mental model that he could hold in his mind, like the mental model he had built of his businesses. This model would help him see how all the different aspects of dyslexia—all its potential strengths and challenges—fit together to make a mind like his. He was asking us to help him build that model.

On one level, creating a coherent model of dyslexia is the primary goal of this book. As you'll discover in later chapters, recent research has shown that one of the core abilities lying beneath all the dyslexic MIND strengths we'll discuss is in fact the ability to make accurate and lifelike models of physical and conceptual worlds. So this will be a distinctly dyslexic kind of task—in a very positive sense. As the book progresses, we'll help you make your own accurate and lifelike model of dyslexia. This model won't just *define* dyslexia for you by listing its features. It will help you *understand* dyslexia in a deep and meaningful way. This deeper understanding should enable you to plan, predict, troubleshoot, and innovate in ways that will equip you to take control of your own dyslexic journey, so you can deal with your challenges and build on your strengths.

Preparation: Eliminating Two Simple— but Improper—Models

Before beginning to build our model, let's quickly set aside two common but mistaken models of dyslexia that people often use, so they don't get in anyone's way.

The first improper model is of dyslexia as a *binary distribution*. Binary distributions are yes/no kinds of things. Things you either have or don't have, that either are or are not. Dyslexia is not like that. The traits associated with dyslexia can be present in different individuals in varying degrees, and they seem to be the product of many different genes as well as of certain experiences during development.

The second improper model is of dyslexia as a *normal distribution*. Most of us are familiar with bell-shaped distribution curves. They are used to describe the presence in a population of single traits that can be measured along a single dimension. Like height: Some people are very tall, others very short, and most somewhere in between. However, all people can be measured and placed along the curve according to this single dimensional variable: height. Making comparisons along single dimensions are relatively easy. They're simply a matter of measuring more or less of that one specific thing. However, dyslexia is not a single trait that can be measured along one dimension. Instead, dyslexia encompasses a mixture of traits, some of which are strengths and some challenges. Even if we look only at the challenge side of dyslexia, there are at least six different traits that have been found to be independent risk factors for dyslexic challenges.

So if dyslexia can't be pictured either as a binary or a normal distribution, how then can it best be pictured? The short answer is that it requires a multidimensional model, one that can express the independent contributions of dyslexia's many different traits.

This is where things begin to get a bit tricky. Most people can easily imagine three dimensions. That's how we think about space. Those interested in physics or science fiction can also easily add the fourth dimension of time. Yet when we talk about five or six or ten or twenty dimensions, most people find themselves at a loss. Fortunately, we actually all think multidimensionally all the time. We just don't realize it. Here's what we mean.

Modeling Dyslexia in Multiple Dimensions

While normal distributions aren't good at describing *dyslexia as a whole*—because dyslexia is not a single trait that can be measured along a single dimension—they *are* good at describing each of the separate component traits (or dimensions) that make up dyslexia. In other words, each of the traits that make up dyslexia can be thought of as being normally distributed along its own unique dimension. So the process of thinking multidimensionally about dyslexia really just means combining into one large model all our thoughts about the whole series of dimensional traits that make up dyslexia. This may sound a bit abstract at first, but it will seem much simpler if we use another model to illustrate.

The model we're going to use is that of a mountain, and the particular kind of mountain we have in mind is a big conical volcano that stands out from all the surrounding mountains. We live near Seattle, so we're going to think about Mount Rainier. If you're more familiar with Mount Fuji or Kilimanjaro or some other similar mountain, then picture one of those.

Why a mountain? Because thriving with dyslexia is like successfully climbing a mountain in several ways. First, the conditions you might face can show an almost infinite variety. Think about the mountain. If you told a friend that you were "climbing Mount Rainier," that would give your friend only a very limited idea of your current whereabouts and condition. To give your friend a really clear idea about your condition—about the challenges you face, your chances of reaching the summit, and your present needs—you would have to provide more information.

It's the same with dyslexia. When you tell someone—like a teacher or a boss—that you are dyslexic, that doesn't say much about your strengths and challenges, or what you will need to reach peak performance. To help those people understand, you need to give them more information.

Let's return to our mountain model. Instead of just telling someone that we are climbing Mount Rainier, we can provide them with the following dimensional variables:

- Latitude
- Longitude
- Altitude

- Time of day
- Time of sunrise and sunset
- Air temperature
- Wind speed
- Luminosity (brightness)
- Visibility
- Precipitation
- Barometric pressure
- Ground traction
- Our physical condition
- Our climbing ability
- Our supplies and equipment

Each variable has its own unique dimension on which it can vary from a greater to a lesser degree. The combination of all these variables creates a rich fifteen-dimensional model, and providing a description of where things stand with each of those variables makes understanding and planning much simpler and more effective.

This example provides a helpful image of what it means to think multidimensionally about a complex concept like dyslexia. It should also help us picture—or *mentally simulate*—how altering our position along any of these dimensional variables could dramatically change our situation on the mountain, requiring different plans or actions. Think, for example, of how radically our situation could change if the air was greatly warmer or cooler; if the time was closer or further from sunset; if the weather was clear, cloudy, rainy, or snowy; if the slope was steeper or flatter, or the ground rougher or more even; if we were rested, physically fit, and well nourished, or sleep deprived, hungry, and out of shape. Climbing Mount Rainier can encompass any of these conditions, and one trip up the same mountain can differ dramatically from another.

Second, although keeping track of this multidimensionality is complex, it is not endlessly complex; modeling even a limited number of key dimensional variables will help us greatly to understand and manage our situation.

We can use precisely this kind of multidimensional approach to understand and reach peak performance with dyslexia. In the rest of this book, we will help you identify the dimensional variables that will form your model of dyslexia. First, we'll consider the neurobiological basis for the dyslexic processing style.

Part 2

The Scientific Dimensions of Dyslexia

4

Sources of Dyslexia's Dual Nature

One of the biggest challenges in identifying the dimensions of our dyslexia model is that dyslexia results from human brain function, and human brain function can be viewed and explained on four different levels. Those levels are:

1. The brain *structural* level, which deals with how brain cells (or *neurons*) are organized into various structures (e.g., hemispheres, lobes, cortex, cerebellum, etc.).

2. The brain *connection* or *network* level, which deals with how neurons link together to form functional circuits.

3. The brain *cognitive* level, which deals with the basic mental tasks or processing functions that are performed by brain networks and circuits.

4. The *behavioral* level, which deals with the visible, real-world manifestations of the brain's cognitive processes—both strengths and weaknesses.

To build our model of dyslexia, we'll need to look at the brain and its functions at each of these four levels. We'll begin in this chapter by focusing on two dyslexic brain differences that can best be understood at the structural and network levels. In chapter 5, we'll examine another key brain difference that is best understood at the network and cognitive levels. Each of these differences helps to explain the strangely dual nature of dyslexia.

One Brain, Two (Very Different) Halves

In 1981, Dr. Roger Wolcott Sperry was awarded the Nobel Prize for his discovery that the brain's two halves, or hemispheres, process information in very different ways. Ever since, a steady stream of books and articles have popularized the idea that there are distinctive right-brain and left-brain thinking styles, and that individuals can be primarily right-brained or left-brained in their cognitive approach. While most discussions of this topic have been oversimplified, in many respects the evidence is now stronger than ever that the brain's two hemispheres really do process information in very different ways.

As a rough generalization, the brain's left hemisphere specializes in fine-detail processing. It carefully examines the component pieces of objects and ideas, precisely characterizes them, and helps to distinguish them from one another. The right hemisphere specializes in processing the large-scale, big-picture, or global features of objects or ideas. It's especially good at spotting connections that tie things together; at seeing distant similarities or relationships between objects or ideas; at perceiving how parts relate to wholes; at determining the essence, gist, or purpose of a thing or idea; and at identifying and taking account of any background or context that might be relevant for understanding the objects under inspection.

We can roughly summarize the differences between left and right hemispheres at the functional or *cognitive level* by saying that they specialize respectively in trees and forest, fine and coarse features, text and context, or parts and wholes. These differences show up in important ways in the brain's various processing systems. Consider vision: When viewing an object, the left hemisphere perceives fine details and component features, but it's poor at binding those features together to see the larger whole. For example, the left hemisphere can recognize eyes and ears and noses and mouths, but it's poor at recognizing faces. Similarly, it can see windows and doors and chimneys and shingles, but it's poor at comprehending houses. To perceive these larger patterns, the left hemisphere requires big-picture processing help from the right.

Several kinds of evidence suggest that at the *network level,* individuals with dyslexia differ from nondyslexics in the ways they use their brain hemispheres to process information. In particular, a growing body of research suggests that for many processing tasks, individuals with dyslexia use processing centers in their right hemispheres more extensively than do nondyslexics. Differences of this type have been shown for many auditory, visual, motor, language, memory, and other cognitive functions. Some of these differences play important roles in reading.

Differences between dyslexic and nondyslexic brains have also been found at the *structural level.* In most individuals with dyslexia, the two brain hemispheres are equal in size, while in nondyslexics the left hemisphere is usually larger than the right.

Several prominent writers have also observed that many individuals with dyslexia show a distinctly right-brained style or cognitive flavor in the ways that they process information. A particularly strong case for this connection has been made by author Thomas G. West in his marvelous book *In the Mind's Eye*. West—who himself is dyslexic—suggests that this right-sided processing pattern may be directly related to the visual and spatial talents shown by many individuals with dyslexia.

Scientists have also found that individuals with dyslexia use their right hemisphere networks more extensively for reading than do skilled nondyslexic readers. This difference was first demonstrated in the late 1990s by Drs. Sally and Bennett Shaywitz at Yale, who used a brain scanning technique called functional magnetic resonance imaging (fMRI) to identify the brain areas that become active as individuals read. Reading expert Dr. Maryanne Wolf summed up the results of this work in this way: "The dyslexic brain consistently employs more right-hemisphere structures [for reading and its component processing activities] than left-hemisphere structures."

While this increased right-hemispheric processing may at first appear to involve a rightward shift from the normal left-sided pattern, it actually reflects the *absence* of the usual leftward shift that occurs as individuals learn to read. Dr. Guinevere Eden and her colleagues at Georgetown University have shown that most beginning readers use *both* sides of their brain quite heavily—just like adults with dyslexia do. It's only with practice that most readers gradually shift to employ primarily left-sided processing structures.

Individuals with dyslexia have a much harder time making this shift to primarily left-sided or expert processing during reading. Without intensive training, they tend to retain pathways that more closely resemble the immature or beginner pathway, with its heavy reliance on right-hemispheric processing.

Importantly, this greater use of right-sided processing structures by dyslexic individuals is not unique to reading but is seen with many cognitive processes, especially ones that become automatic and highly efficient with practice. This dyslexic tendency to retain more right-sided processing raises two important questions. First, *why* do individuals with dyslexia show this persistence of heavy right-hemisphere involvement? And second, what are the *consequences* of this persistence for dyslexic thinking and processing?

Let's consider the first question. The practice-induced right-to-left processing shift generally seems to accommodate changes in our processing needs as our skills with repetitive procedural tasks increase. When we first attempt a new task, our right hemisphere's big-picture processing helps us recognize the overall point or essence of the task, so we don't get lost in the details. It also helps us recognize how the new task may be similar to tasks we've learned before. This helps us problem-solve and fill in details we miss. In these ways, the right hemisphere's top-down or big-picture processing focus is ideal for our early awkward attempts to stumble through processes we're still fuzzy on.

As we become more familiar with the purposes and demands of a task, our need for big-picture processing diminishes and is replaced by a need for greater accuracy, automaticity, and speed. That's where the left hemisphere comes in, with its greater ability to process the fine details that must be mastered to develop automaticity and expertise.

One well-documented example of a processing shift from the right to the left hemisphere that occurs with training is the shift that takes place in expert musicians. Researchers have shown that untrained music listeners process melodies primarily with their right hemispheres, so they can grasp the large-scale features (or gist) of the melody. By contrast, expert musicians process music more heavily with their left hemispheres, because they focus on the fine details and technical aspects of the performance.

This tendency to shift from right- to left-hemisphere processing as automatic skills increase through practice is intriguing, because it suggests that the delays seen in dyslexic individuals to make such shifts might reflect a kind of general difficulty in acquiring automaticity through practice. This is in fact the case: Dyslexic individuals have indeed been found to be slower in developing automaticity for all sorts of rule-based skills and procedures, including those involved in reading, as we'll discuss in more detail in chapter 5 when we look at the *behavioral level*. Such delays in mastering rule-based skills could clearly slow the development of "expert," left-sided pathways and cause prolonged dependence on the "novice," right-sided circuits, both for reading and for the other processing tasks we mentioned.

If delays in developing automaticity help to explain why individuals with dyslexia rely more on their right hemispheres for many tasks, then what are the consequences of this more right-hemispheric processing style? We can begin to answer this question by looking at cognitive-level differences in how the right and left hemispheres process information. Let's use language as an example.

In 2005, Northwestern University psychologist Dr. Mark Beeman published a fascinating paper describing the differences in the ways that the two brain hemispheres process language. When the human brain is presented with a particular word, each hemisphere analyzes the word by activating its own *semantic field*, or collection of definitions and examples describing that word. Importantly, the semantic fields contained in the left and right hemispheres perform this analysis in significantly different ways.

The left hemisphere activates a relatively narrow field of information, which focuses on the primary (or most common and often most literal) meaning of the word. This narrow field of meaning is particularly well suited for processing language that's low in complexity or requires precise and rapid interpretation—like comprehending straightforward messages or following simple instructions. It's also useful for quickly and efficiently *producing* language. Since speaking and writing require the rapid production of *specific words* (rather than blended or compound words), the less ambiguity or hesitation, the better. The left hemisphere's narrow semantic fields are ideal for such production.

The right hemisphere, by contrast, activates a much broader field of potential meanings. These meanings include secondary (or more distant) word definitions and relationships, like synonyms and antonyms, figurative meanings, humorous connections, ironic meanings, examples or cases of how the word can be used or what it represents, and words with similar styles (e.g., formal/informal, modern/archaic) or themes (e.g., relating to the beach, to chemistry, to emotions, to economics). This broader pattern of activation is slower, but it's also much richer. That's why it's particularly useful for interpreting messages that are ambiguous, complex, or figurative. Tasks for which the right hemisphere is particularly helpful include comprehending or producing metaphors, jokes, inferences, stories, social language, ambiguities, or inconsistencies.

We asked Mark to illustrate the kind of distant connection that right hemispheric semantic processing is particularly good at detecting. He responded with the following example. "Consider this sentence: 'Samantha was walking on the beach in bare feet, not knowing there was glass nearby. Then she felt pain and called the lifeguard for help.' When most people hear that sentence, they infer that Samantha cut her foot. But notice that the sentence never explicitly states that she cut her foot, or even that she stepped on the glass. These facts have to be inferred, and these inferences are made by the right hemisphere. It produces these inferences by detecting the overlap in semantic fields between the terms *bare feet*, *glass*, and *pain*."

The right hemisphere's special skill in making such distant connections and inferences is just what individuals with dyslexia need when reading and listening. Decoding problems often make it hard for individuals with dyslexia to identify printed words, and problems distinguishing closely related words can cause similar problems with listening. As a result, individuals with dyslexia must often use contextual clues to fill in parts of messages they've missed. This is precisely what the right hemisphere excels at. Rather than being a source of the reading or listening problems of dyslexic individuals, the dyslexia-associated increase in right-hemisphere processing may actually be an ideal compensation for individuals who are struggling to process language at its most basic levels.

The dyslexia-associated bias toward right-hemisphere processing may also help to explain the unusual mix of challenges and strengths that is commonly seen in individuals with dyslexia. In his research describing semantic fields, Mark showed that the right hemisphere's broader conceptual- or cognitive-level fields were mirrored at the brain's structural and network levels by physically broader and longer-range connections between neurons. On the challenge side, these physically broader and more diffuse connections in the right hemisphere can lead to slower, less efficient, less accurate, and more effortful processing. However, the broader network of connections provided by the right hemisphere may also favor new and creative connections, the recognition of more distant and unusual relationships, and skill in detecting inferences and ambiguities.

If we pause to think about it, strengths of this sort are not typically the hallmarks of a beginner, but of an expert. This raises an important point about expertise: It comes in many forms. Some forms of automatic expertise enable fast, accurate, and unthinking performance of repetitive tasks. Other more mindful and conscious forms equip us to think through and solve challenging problems. This is where the allocation of brain resources gets interesting, because novelty—that is, tasks we must approach like beginners—is not the only type of task that leads to greater involvement of the right hemisphere. Particularly difficult tasks, including those that force us to take account of complex contexts and conditions, or to draw on past experiences, analogies, or examples, rather than rules and formulas, also more heavily engage the right hemisphere.

In a fascinating study from Canada, attending physicians and medical students were presented with difficult clinical problems to solve while their brain activity was monitored in a functional MRI scanner. The medical students preferentially used areas in the brain's left hemisphere that are associated with linguistic reasoning and rules and definitions to solve the difficult problems. In contrast, the experienced senior physicians preferentially used areas in the right hemisphere that are associated with personal experience, imagination, creativity, and performing mental simulations. As we'll see in later chapters, these areas overlap highly with the brain regions that dyslexic individuals commonly use when they engage their MIND strengths. Rather than seeing the persistence of right-sided processing as a sign of failure to develop expertise, it might be more accurate to see it as a bias toward using one form of expert processing in preference to another.

Alterations in Microcircuitry: Big-Picture Versus Fine-Detail Processing

A second important difference that distinguishes dyslexic and nondyslexic brains was first described by Dr. Manuel Casanova, who is currently a professor of biomedical sciences at the University of South Carolina School of Medicine. In pursuing his broad interests as a psychiatrist, neurologist, and neuropathologist, Manny has studied many different "types" of brains, including those of individuals with dyslexia and autism. He is particularly interested in structural- and network-level differences he has found in the ways the neurons or nerve cells are organized and linked with one another in the part of the brain known as the *cortex*. Before we explain his findings and how they may relate to dyslexic advantages, let's take just a moment to review a few things about the structure and function of the cortex.

At the structural level, the cortex is a thin sheet of cells that coats much of the brain's surface. This thin sheet contains six even thinner layers of cells, each of which performs different functions and often contains subtly different kinds of neurons. The neurons in the cortex communicate with one another using a combination of chemical and electrical signals. Through the process of exchanging these signals, they give rise to our higher cognitive functions, like memory, language, sensation, attention, and conscious awareness.

Although different parts of the cortex typically perform different functions, they are all very similar in the ways that they are organized and perform their tasks. In every part of the cortex, the cells in the six cortical layers are organized into functional units called *columns*, which are in turn made up of smaller functional units called *minicolumns*. These minicolumns will be our primary focus in this section.

Minicolumns were first discovered by researchers who inserted tiny electrodes into the brain's cortex to record its electrical activity. When they pushed their electrodes perpendicularly into the surface of the cortex, like a candle into a birthday cake, they found that the cells stacked directly on top of each other in all six cortical layers responded to stimuli in unison. In contrast, when they inserted the electrode at an angle to the brain's surface, the cells in the different layers did not fire together. These results suggested that the cells in the six cortical layers were grouped functionally into tiny cylindrical units running perpendicular to the surface of the brain: hence, minicolumns.

Minicolumns are the basic information processing units in the brain's cortex. The reason they look so similar in different parts of the brain is that they all perform very similar functions. As pioneering technologist and neuroscientist Jeff Hawkins explains in his book *A Thousand Brains,* "What makes [these different processing units] different is not their intrinsic function but what they are connected to. If you connect a cortical region to eyes, you get vision; if you connect the same cortical region to ears, you get hearing; and if you connect regions to other regions, you get higher thought, such as language. . . . [T]hroughout the neocortex, columns and minicolumns perform the same function: implementing a fundamental algorithm that is responsible for every aspect of perception and intelligence."

In other words, to process more than just the most basic kinds of information, minicolumns must be linked to form circuits, just as the microchips in your computer must be linked to create complex processing functions. But instead of being soldered together, minicolumns are connected by long cellular projections called *axons,* which stretch like cables from the neurons in one minicolumn to the neurons in others.

When Manny examined the structural arrangements of minicolumns in different brains, he found that each brain had its own characteristic pattern of spacing between minicolumns. Some brains contained tightly packed minicolumns, while in other brains the minicolumns were more loosely spaced. In the human population as a whole, minicolumn spacing forms a bell-shaped or normal distribution. In most people this spacing is close to an average value, while smaller numbers of people have minicolumns that are either more widely or more tightly spaced. When Manny looked more closely at the "tails" of the distribution, he found something interesting. As he told us, "When we looked at the end of the spectrum that was characterized by widely spaced minicolumns, we found a very high proportion of individuals with dyslexia. Not surprisingly, when we looked, we found a high proportion of individuals with autism in the other tail of minicolumn spacing, where the minicolumns are closely packed."

Manny also noticed an interesting relationship between minicolumn spacing and the size and length of the axons that connected the minicolumns. He found that brains containing tightly packed minicolumns sent out shorter axons, and that they formed physically smaller or more local circuits. In contrast, brains with more widely spaced minicolumns sent out larger axons that formed physically longer-distance connections, creating larger circuits and involving more distant parts of the brain.

This bias at the network level toward forming either longer-range or local connections has important cognitive-level consequences, because the circuits formed by these different connection types excel at different tasks. Local connections are especially good at processing fine details—at sorting and distinguishing closely related things, like different sounds, sights, or concepts.

Brains biased toward forming these shorter connections generally show a high level of skill in detail-oriented tasks that involve extracting or identifying the precise features of objects or ideas.

Longer connections, like those enriched in dyslexic brains, are poorer at fine-detail processing but excel at recognizing large features or concepts—that is, at big-picture tasks. Examples of such big-picture tasks include recognizing the overall form, context, or purpose of a thing or idea; synthesizing objects and ideas; perceiving relationships; and making unusual but insightful connections. Circuits formed from such long connections are also useful for tasks that require problem-solving—especially in new or changing circumstances—though they are slower, less efficient, and less reliable for familiar tasks and less skilled in discriminating fine details.

Let's pause for a moment to reflect on how closely the strengths associated with short or local connections match the kinds of left-brain processing skills we discussed in the previous section. Then note how closely the strengths associated with long or distant connections also match the previously mentioned right-brain processing skills. Finally, notice how closely the processing style associated with longer connections—that is, the "strong big-picture/weak fine-detail" trade-off—matches the cognitive style we've described as being common among individuals with dyslexia. In essence, not only do dyslexic individuals tend to use their right hemispheres more extensively for many processing tasks, as we saw earlier, but due to their wider minicolumn spacing, even the neurons in their left hemispheres tend to form broader networks of connections that more closely resemble the typical right-brain pattern.

We asked Manny to explain in simple terms why a bias toward longer connections might favor big-picture processing. He began by explaining that higher cognitive skills arise when minicolumns are connected to form *modular systems*. He illustrated what this would mean by using the example of a car. "Cars have many separate components, or modules, such as the transmission, motor, and tires. When these modules are connected into a larger system, they create new or *emergent properties,* like the property of locomotion. This property isn't present in any of the separate modules—it emerges only when the modules are connected to form a larger system or network. This example shows how the properties of the modular system as a whole can greatly exceed the properties of its individual elements, and how connecting these elements in particular ways can create new functions that wouldn't exist if the parts weren't connected or were connected in some other way. In the brain it is similar. Depending on how you link minicolumns together into circuits or networks, you get the emergence of higher cognitive functions, like judgment, intellect, memory, orientation. Those functions weren't there within the properties of the individual minicolumns. They emerged as the appropriate connections were made between cells in different parts of the brain. In other words, broader connections favor the formation of broadly integrated circuits, which in turn create high-level cognitive skills."

Because minicolumns are involved in essentially all higher cognitive functions, we might expect that factors affecting the arrangements of an individual's minicolumns at the structural and network levels—such as the ways they are spaced and connected—might create important functional differences for many of that individual's higher brain functions. Again, this is just what we do see in individuals with dyslexia. As a group, they differ from nondyslexics not only in how they process printed words or spoken language, but also in how they perform almost every type of higher brain function, including memory, sensory and motor tasks, attention, and language.

According to Manny, the bias shown by dyslexic brains toward forming long-distance connections leads to the emergence of *both* strengths and challenges in many areas of function—that is, both to big-picture processing skills *and* to weaknesses in fine-detail processing. Manny cited phonological processing, or the processing of the basic sound units that are combined to create words, as one example of a fine-detail processing task that is often particularly hard for individuals with dyslexia. In fact, their difficulties in this area play a key role in creating their dyslexia-associated challenges with reading and spelling. Difficulties with fine-detail processing may also explain many of the other dyslexia-associated challenges associated with listening, vision, motor function, and attention.

Manny contrasted the difficulties that individuals with dyslexia experience in fine-detail processing with the strengths shown in these tasks by many individuals with autism. "The brains of individuals with autism are biased toward short connections at the expense of long connections—just the opposite of dyslexia. Cognitively, individuals with autism focus on particular details: they see the trees but lose the forest. If you test patients with autism, their thinking tends to be rather concrete, and they struggle to see the broader meaning, form, or context. However, where they often excel is at tasks that can be performed using a tightly localized brain region, because they require only one specific function. An example would be finding Waldo in the *Where's Waldo?* books. This fine-detail processing task is performed entirely within one highly localized area of the visual cortex, where the tightly packed minicolumns are connected by many short axons into a local circuit that excels in fine-detail processing. With such tasks, individuals with autism often perform much better than other people.

"On the other hand, where individuals with autism often

struggle is with tasks like face recognition, which require that many different processing centers spread all around the brain work together. This joining or 'binding' of distant processing centers is very hard for individuals with autism, because they don't easily form the necessary long-distance connections.

"In contrast, joining distant areas of the brain together is just what individuals with dyslexia do best. As a result, individuals with dyslexia excel at drawing ideas from anything and anywhere, and at connecting different concepts together. Where they may miss the boat is in processing fine details."

The most intriguing aspect of these findings is how well the predicted results of this single brain variation match with the actually observed patterns of challenges and strengths found in individuals with dyslexia. This correlation argues strongly in support of its importance.

Conclusion

In this chapter we've focused primarily on the brain's structural and network levels to understand two key brain differences commonly observed in individuals with dyslexia: a preference for right-brain processing and a bias for widely spaced minicolumns. Both differences contribute to the duality of strengths and challenges that is dyslexia's most remarkable feature. In the next chapter we'll examine a third key difference.

5

More Conscious, Less Automatic

The third important brain difference distinguishing dyslexic and nondyslexic brains is the tendency shown by dyslexic brains to perform more processing tasks using *conscious* rather than *automatic* mental resources. This is a topic of enormous importance for understanding dyslexic strengths and challenges, and we'll need to look at all four of the levels mentioned in chapter 4 to fully explain it. As we'll see, some of the same underlying factors that lead to greater right-hemispheric processing in dyslexia are also involved.

Two of the researchers who have done most to reveal this important difference are Drs. Angela Fawcett, professor emeritus of psychology at Swansea University in Wales, and her longtime collaborator Roderick Nicolson, professor of psychology at Edge Hill University in England. Together, Angela and Rod have proposed a detailed model of dyslexic brain development that not only summarizes the differences in automatic processing that distinguish dyslexic and nondyslexic learners, but also explains how those differences may arise during brain development.

Before examining this model in detail, we'll have Angela provide some key background on a cognitive-level system that plays a crucial role in learning how to perform tasks automatically—a process called *procedural learning*.

Procedural Learning and Memory

Angela described procedural learning for us in the following way: "Procedural learning is learning *how* to do something, and learning it to the point where it's automatic, so you know how to do it without having to think about it. This process of becoming automatic with complex rules and procedures is much more difficult if you're dyslexic."

At least half of all individuals with dyslexia can be shown on cognitive testing to have significant problems with procedural learning. In addition, as we'll see in later chapters, a large majority of dyslexic individuals acknowledge difficulties performing procedural tasks.

As a result of these differences in procedural learning, dyslexic individuals are usually slower to master any rule-based, procedural, or rote skill that should become automatic through practice. Because most basic academic skills are heavily rule and procedure dependent, problems with procedural learning can cause a wide range of academic challenges. These are often especially intense in the early grades. We'll discuss some of the specific academic difficulties associated with procedural learning challenges in the next chapter.

Dyslexic individuals with procedural learning challenges struggle more than most nondyslexic people to learn to perform tasks simply by observing others perform those tasks and then imitating them—a process known as *implicit learning*. Instead, they learn better when rules and procedures are demonstrated and explained clearly, a process known as *explicit learning*. Since individuals who struggle with procedural learning have difficulty performing rule-based skills automatically, they must instead use *conscious compensation*—that is, they must use information they've learned explicitly to consciously talk themselves through tasks.

As we'll see later, there are benefits to this need to think explicitly about tasks, but there are also important drawbacks. This is especially true during the early stages of learning, when many tasks have not yet been fully mastered. The biggest drawback is that conscious compensation places a heavy load on working memory. To understand the significance of this burden, let's take a moment to better understand working memory.

Working memory is the form of short- to intermediate-term memory that helps us keep things in mind for active conscious processing. It's the brain's version of the random-access memory or RAM on your computer that allows you to run the programs you have open. The form of working memory that is especially relevant to dyslexia is called *auditory-verbal working memory*. Auditory-verbal working memory helps you keep word- or sound-based information active or in mind until it can be processed and put to use. It's the mental space where you hold discussions with yourself while reasoning and making decisions. It's also where you repeat instructions to yourself while you work through an unfamiliar or complex task. Working memory is especially important for the tasks you are still learning to perform and haven't yet mastered to the point where you can perform them automatically (that is, without thinking about them). During complex tasks, working memory resources can easily become overwhelmed, and mistakes may result. Because individuals with procedural memory problems must perform so many tasks using conscious processing, they will often experience this type of *working memory overload*. This makes them slower and more error-prone than others on routine tasks.

Individuals who struggle with procedural learning also require many more repetitions than others to master complex skills. As Angela explained: "You can teach a dyslexic child what the rules are, and she appears to grasp them, but then the rules slip away again. We actually came up with something we called the *cube root rule*, which means that it takes the cube root *longer* to learn something if you're dyslexic than if you aren't. In other words, if it took eight hours to learn something for a nondyslexic, it would take twice as long for a dyslexic [2 being the cube root of 8]; and if it took one thousand hours, it would take ten times as long, or ten thousand hours [10 being the cube root of 1,000]. So you can see how much extra work is needed to get these children to develop skills similar to other children."

Finally, individuals with dyslexia and procedural learning challenges also tend to forget skills they appear to have mastered more quickly than others if they don't regularly practice them. According to Angela, "Often teachers will say, 'This child seemed to have learned this before the six weeks' holiday, but now he's come back and it's gone.' It helps to show teachers that it's not due to a moral fault in the student or any lack of effort. Actually, the dyslexic child is working much harder than everybody else, and this difficulty in learning and retaining rules results from a fundamental difference in the learning process. When you understand this, you realize that it's not something the child should be ashamed of, but something he should be taught to get around, using specific strategies."

With this basic understanding of automaticity and procedural learning in mind, we're now ready to explore how dyslexic differences in these areas might develop, and what further consequences they might have for dyslexic strengths and challenges.

Neural Noise and Delayed Network Formation

One of the key elements in Rod and Angela's model of dyslexic brain development is the finding that at the neural connection or network level, signals sent between nerve cells (*neurons*) in dyslexic brains typically show greater than normal levels of *neural noise*. Neural noise refers to the greater variability in the clarity of the signals that neurons transmit across their cell-to-cell connections (called *synapses*) in response to inputs like sights or sounds. This variability in the *signal-to-noise ratio* can cause several problems for the brain, but most important, it makes it very hard for the procedural learning system to learn to recognize and respond to particular stimuli in a fast, automatic, and consistent manner.

Think, for example, of a teacher speaking to her class who pronounces a word sound associated with a particular letter, then repeats it several times. Now think of a student in that class whose auditory neurons are especially noisy or variable in the firing patterns they generate in response to the teacher's speech. That student's brain may respond as if it heard several different stimuli rather than an identical stimulus repeated several times. That kind of variability in processing can have a big impact on procedural learning. The most fundamental principle of brain-based learning is called Hebb's rule, which states that the neurons that fire together wire together. Neural noise can disrupt the generation of consistent co-firing in response to similar inputs. That's why the failure of connected neurons to respond consistently to similar stimuli will hinder learning from occurring and prevent the development of fast, accurate, and automatic responses to that stimulus.

Rod Nicolson summed up the nature of this problem for us in the following way: "This sort of noise means that the cerebellum, which is a key structure in learning automatic responses, is not getting a consistent picture across time from which it can develop an accurate impression of the input. Therefore it cannot work as efficiently as it should. When the signals are even more variable, it will have a big disruptive effect all down the line."

In addition to this variability in signal-to-noise ratio, the neurons in dyslexic brains also show higher response thresholds to various kinds of inputs when compared with neurons in non-dyslexic brains. As one example of the kind of difficulty that can result from these higher thresholds, it has been shown that dyslexic readers' eyes will commonly drift further off a fixation point before their brains perceive and halt that motion by firing off motion detection cells. This higher threshold for motion detection makes it harder for dyslexic readers to fix their eyes on a single spot on a page, and it results in unstable eye movements that can cause letters to be poorly registered by the retina, blur, or appear to shift places. These and other challenges in correctly registering fine visual details are key contributors to the reading challenges of many dyslexic students.

A similar process takes place with hearing, where very rapid or subtle sound distinctions may fall below the threshold of detection. This higher response threshold may make it harder to distinguish the sounds in certain words, like *time* and *dime*, or *van* and *fan*. Threshold differences like these have been shown to contribute to dyslexia-related difficulties in processing word sounds, and in developing the automatic skills that rely on such information. These difficulties in *phonological processing* are one of the largest contributors to dyslexic decoding and spelling challenges, particularly in English.

Rod and Angela have proposed several possible sources for the higher response thresholds and greater variability shown by the neurons in dyslexic brains. One of these sources is the structural-level pattern of broader minicolumn spacing that we discussed in chapter 4. Longer distances between neighboring neurons have been found to decrease the number of synaptic connections that form between them. This raises the threshold for signal transmission from neuron to neuron. Longer connections can also decrease the strength of transmitted signals, resulting in failure to reach the thresholds necessary for signal transmission.

Whatever the source of this signal variability, it impairs the formation of the brain networks needed to perform complex tasks—like reading fluently. Rod explains the importance of this network formation in the following way. "A lot of older work on the brain structures underlying reading skills tended to be localist, stating that 'this part of the brain does this sort of processing.' It was only more recently that people were able to look at functional imaging and to see which parts of the brain were involved when people were actively performing tasks. That imaging allowed us to see how different parts of the brain are joined to each other to form into functional networks that actually perform these complex processes. These networks develop as the child develops, and they can take quite a long time to mature. This realization has caused Angela and me to expand the way we have looked at the sources of dyslexia. We began with the automatization deficit hypothesis: the idea that dyslexic children take longer to make their skills automatic, so they can do them without having to think about it. Now we expand to this deeper network level and say that it takes dyslexic children longer to develop these more efficient structural networks as well."

To recap briefly, dyslexic brains show greater variability in signal clarity and face higher activation thresholds when transmitting signals between neurons. Both differences contribute to difficulties automating routine, habitual, procedural skills by slowing the formation of the efficient brain networks needed to perform these complex tasks.

Automaticity: Triumph or Trade-Off

While these consequences may at first appear grim, on closer inspection we may begin to see that efficient nerve cell communication, automatization, and network formation aren't simply the unmixed blessings that we thought them at first glance. They may produce their own kinds of two-sided trade-offs.

Consider, for example, our discussion in the last chapter of the way that continued practice typically leads to a right-to-left shift in hemispheric processing for many repetitive tasks. This shift typically enhances efficiency, but at a cost to holistic or gestalt processing. This raises a question: Might a delay in efficient network formation in dyslexic children—and the continued reliance on more holistic right-hemisphere processing that results—be a potential source of cognitive strengths?

Rod and Angela have actually raised this question. They suggest that the delayed commitment that neurons show in forming efficient networks may create cognitive benefits. Rod told us: "I think that in many cases it is useful to take more time to develop your networks, because the side effect of developing an efficient network is that you can lose other skills and knowledge that you had before. It's a bit like what happens when you create a superhighway between places A and B. You no longer go through all the towns en route. You just travel at speed past those little villages and all of the knowledge they contain. So it may well be that the consequence of our building these superhighway networks in the brain is that we no longer routinely encounter the initial building blocks of our knowledge, and therefore we have fewer chances to integrate them in other and richer ways. Because in order to integrate two ideas or concepts in your mind, you need to have both of them activated at the same time. When you become more efficient and automated, you classify and pigeonhole all the knowledge and experiences that initially formed your skills and higher concepts, and because you don't have to think about those sources again, you lose the ability to use that knowledge to synthesize new ideas. That to my mind is one of the key benefits to what we've called *delayed neural commitment,* or the tendency to take longer to create efficient networks. Because you haven't been quite so efficient at analyzing and automating all this stuff—that is, at building your superhighway networks—you're still able to do a fair bit of synthesis, and you can keep that synthesis route open a bit longer in life. So it might make you very much more creative."

Declarative Memory: Keeping Memory More Conscious

In this last statement, Rod is touching on something that Angela mentioned in her description of procedural learning: Because dyslexic individuals are less efficient in automating skills and making their performance unconscious or *implicit*, they must perform more tasks using conscious, aware, or *explicit* processing. While automatic or implicit programs are stored in *procedural memory*, explicit instructions are stored in the other major branch of long-term memory, which is called *declarative memory*.

Declarative memory stores memories that can be consciously recalled and put into words, or declared to your conscious mind in an explicit and comprehensible form.

This greater reliance on declarative memories is an asset for individuals with dyslexia. If we employ the language from Rod's analogy, dyslexic individuals make more of their mental journeys on local roads than superhighways, so the local details are more readily in sight. As a result, these less automatic dyslexic processors will have more opportunities to synthesize and combine basic information in new ways. Their responses may be slower and less efficient, but they will also be less routine and may be more thoughtful, fresher, and more innovative.

This tendency to use conscious or declarative memory instead of unconscious or implicit procedural memory will also turn out to have important implications for the MIND strengths we'll describe in coming chapters. In fact, during the decade since *The Dyslexic Advantage* was first published, advances in research have convinced us that differences in the ways that dyslexic minds process long-term memories play a central role in each of the MIND strengths and unite them in important and exciting ways.

Conclusion

In the past two chapters we've focused on several key ways in which dyslexic and nondyslexic brains differ at the structural, network, and cognitive levels. These differences include:

- greater relative right-hemisphere size, and a tendency to use right-hemispheric processing centers more extensively for all sorts of tasks

- greater spacing between cortical minicolumns

- increased reliance on declarative or conscious memory, instead of automatic or procedural memory

We believe that these differences are important contributors to the dual nature of dyslexic strengths and challenges. Before we begin our detailed discussion of the MIND strengths that arise from these differences, let's briefly look in the next chapter at some of the dyslexia-associated challenges that also arise from them, because these challenges will also form important dimensions in our dyslexia model.

6

Dyslexia-Associated Challenges

In this chapter we'll review six cognitive-level processes that have each been shown to contribute independently to the likelihood that an individual will experience dyslexia-related challenges with reading and spelling.* Our goal in discussing these processes is to provide you with enough practical information so you can:

- understand and recognize how these processes contribute to dyslexic challenges;

- incorporate your understanding of these processes into your multidimensional model of dyslexia; and

- use the resulting model to understand, troubleshoot, and develop practical ways to deal with any difficulties created

---

* For those who are interested in a detailed list of the visible signs or behaviors that often point to dyslexia-associated challenges, you can link to a list on our website at dyslexicadvantage.org/dyslexia-signs/.

by these processes so you can thrive at school and work as an individual with dyslexia.

Cognitive Processes Underlying Dyslexic Challenges

Phonological Processing

The first cognitive process we'll consider is *phonological* or word sound *processing*. *Phonemes* are the basic sound units that make up words. Every language has its own set of phonemes. English has about forty-four. Just as the letters in an alphabet can be combined to form printed words, phonemes can be combined to form spoken words.

During the past forty years, problems with word sound processing have been shown to be one of the most common contributors to dyslexic challenges with decoding (sounding out) and spelling words. In fact, problems with word sound processing have been found in about 70 to 80 percent of individuals with dyslexia.

Phonological processing is a good example of the kind of fine-detail processing with which many dyslexic individuals struggle. The precise source of the dyslexia-associated challenges with phonological processing at the deeper brain structural and network levels is not precisely known. However, several of the dyslexic brain features that we mentioned in the last chapters, such as broader minicolumn spacing, increased neural noise and higher signaling thresholds, and challenges with procedural learning, have each been mentioned as possible contributors.

Whatever the source, individuals who struggle with word sound processing typically have problems accurately perceiving the sounds that make up spoken words (a skill called *phonemic awareness*). They also have difficulty analyzing and manipulating word sounds, by switching, adding, or dropping sounds from words (a skill termed *phonological awareness*). These problems with sound perception and manipulation make it hard to decode and spell, because they underlie the ability to accurately match word sounds with the letters needed to represent them.

When severe, problems with word sound processing can lead to difficulties not only with decoding and spelling but also with learning new words, mastering syntax and grammar, and comprehending sentences and paragraphs. When especially severe, problems of this type are called *developmental language disorder*, and children in the preschool and early elementary years who show such problems should receive a comprehensive language assessment.

Naming Speed

The second key cognitive function is known as *rapid automatized naming* (RAN), or *naming speed* for short. Naming speed is pretty much what it sounds like: a measure of how quickly an individual can clearly speak the name of a set of familiar items. Naming speed is commonly measured by having individuals say as quickly as they can—while still doing so clearly—the names of fifty or so objects that are pictured in rows on a page. As simple as this sounds, it is very effective in identifying individuals who are likely to have difficulty learning to read *fluently*—that is, with appropriate speed and accuracy. Why is such a simple test so effective? Because it uses many of the same visual, language, and attention skills that reading itself does. In other words, naming speed depends upon the presence of a well-functioning brain network connecting each of the processing centers that perform every one of these functions. As a result, anything that impairs the formation of such brain networks (which we discussed in the last chapter) can result in problems with naming speed.

Problems with naming speed usually lead to difficulty gaining reading speed or fluency. These problems with naming speed can occur either along with, or independently of, problems with word sound processing. One study found that among individuals with dyslexia, about 20 percent showed problems only with word sound processing; another 20 percent showed problems only with naming speed; and most of the remaining 60 percent showed problems with both. When word sound processing and naming speed difficulties occur together, they have been called the *double deficit*. Students with this combined pattern typically have an especially hard time mastering fluent reading skills.

Visual Processing

The third key cognitive function is *visual processing*. Most dyslexia experts now acknowledge that visual processing problems play a significant role in the reading challenges of many individuals with dyslexia. Visual processing is another example of a fine-detail processing task, so it is not surprising that many dyslexic individuals show challenges in this area.

One of the crucial visual skills involved in reading is *visual attention*. Visual attention—as its name suggests—helps your brain decide what visual information to pay attention to and what to ignore. One way to measure your visual attention is by assessing your *visual span*, or how many printed symbols (like a row of letters or numbers) you can take in at a single glance. Studies have shown that, on average, dyslexic readers have narrower visual spans than do nondyslexic readers.

A narrow visual span can make fluent reading difficult. To read fluently, your eyes must perform a complex dance that involves alternating movements, called *saccades*, separated by intervening pauses, called *fixations*. To be effective, a saccade must travel just the right distance relative to the visual span—neither so far that you miss clusters of letters, nor so near that you see a lot of overlap. To correctly plan the length of this leap, individuals must divide their attentional focus so they can plan the target for their next saccade while still concentrating on the place where their eyes are currently fixed. Many dyslexic individuals struggle with this process.* Also, when visual span is overly narrow, it may require many more (and shorter) saccades to read a passage of text. It may also make it more difficult to find one's place if a saccade comes to rest outside of the previous visual span.

During each fixation, the visual focus must remain in one place so the eyes can register all the letters that are contained within the visual span. If the eyes are not completely fixed in one place but instead keep moving, this can lead to blurring, wobbling, or blending of letters in a line of text. These problems are worsened when text is crowded due to small fonts, long or tightly packed lines, or certain styles of type. Many dyslexic individuals have difficulty detecting when their eyes continue moving during the fixation periods, and this can lead to challenges with reading accuracy and speed.

Procedural Learning

The fourth key cognitive function is *procedural learning,* which we discussed in chapter 5. While the importance of procedural learning for individuals with dyslexia has often been underappreciated by professionals, dyslexic people typically find that infor-

---

* This process is called *covert orienting.*

mation about procedural learning has an immediate and powerful resonance. In fact, in *The Dyslexic Advantage* first edition Kindle e-book, the sections on procedural learning were the most highlighted of all. That's because the predicted impacts of procedural learning challenges mirror so closely the actual experiences of many dyslexic people.

As we mentioned earlier, individuals with procedural learning challenges are usually slower to master any rule-based, procedural, or rote skill that should become automatic through practice; and because most basic academic skills are heavily rule and procedure dependent, problems with procedural learning can cause a wide range of academic challenges. Most language skills, for example, require the constant, rapid, and effortless application of rules and procedures. Children with procedural learning difficulties may struggle to master any or all these skills. Such language skills include:

- differentiating one word sound from another
- correctly articulating word sounds and correctly pronouncing words
- breaking words down into component sounds
- mastering the rules of phonics underlying reading (decoding) and spelling (encoding)
- recognizing rhymes
- recognizing how changes in the forms of words can change word meanings and functions (morphology; e.g., *run, ran, running, runner, runny*, etc.)
- interpreting how differences in sentence organization and word order can affect sentence meaning (syntax)

- recognizing language style and pragmatics (the language conventions that carry important social cues)

Many other academic skills are also rule based, such as:

- rote (or automatic) memory of things like math facts, dates, titles, terms, or place-names
- memorizing complicated procedures or rules for things like performing long division, carrying over, borrowing, or dealing with fractions in math
- sequences, like the letters in the alphabet, the days of the week, or the months of the year
- writing conventions like punctuation and capitalization
- motor rules for forming letters and spacing evenly between words the same way every time when writing by hand

Working Memory

The fifth key brain processing function is *working memory,* and in particular, auditory-verbal working memory. Lower auditory-verbal working memory capacity is very common in individuals with dyslexia and may be caused by problems with procedural learning or with auditory fine-detail processing associated with higher neural noise and processing thresholds.

In the last chapter we briefly discussed working memory's role in helping us keep things in mind for short periods of time during active conscious processing. Problems with working memory alone will not usually cause dyslexic decoding and spelling problems, though it may impact reading comprehension. Instead, working memory acts primarily as an intensifier—that is, it increases the impact of the other factors we've already discussed in producing dyslexia-related challenges by limiting your ability to talk yourself through tasks that are difficult or that you have not yet mastered. *To read, you must perform many such tasks.* Here's a partial list of the complex skills that are involved in reading a passage of text. You must:

- recognize the letter symbols
- remember how they relate to word sounds
- decide how the letters and sounds work together to form words
- retrieve and consider the information stored in long-term memory about word meaning
- combine single-word information into phrases and sentences using the rules of grammar and syntax
- recognize idioms, figures of speech, and other phrase- or sentence-level information
- combine all this information into larger passages while analyzing style and literal and figurative meanings, applying background knowledge, and performing other higher-order reading functions

It's truly an enormous job, and if you haven't yet mastered any of these skills to the point where you can perform them automatically, you'll have to talk yourself through those steps using working memory.

For beginning and struggling readers, there is often more information to process than their working memories can hold. As a result, their whole processing system freezes up. Such readers may feel like their mind goes blank or hits a wall—like when your computer bogs down or freezes when you try to run too big a program, or traffic slows to a crawl when too many cars try to share the same stretch of road.

There are several important things to remember about working memory. First, working memory challenges often have their greatest impact on readers who have not yet mastered basic reading skills and who must still talk themselves through the steps. However, they can also cause problems at any time when new skills or tasks are introduced or when the complexity of work is high relative to ability. It is not uncommon to find students who read texts composed of familiar words and clear sentences with reasonable speed and comprehension, but who struggle significantly when new words and complex sentences are introduced, or when they move up to a higher level of instruction. In some cases, this may occur at the college or even graduate school level.

Second, working memory capacity is not fixed. It increases gradually and continually throughout childhood and young adulthood. It expands especially rapidly beginning in adolescence and reaches its maximum capacity only around age twenty-five. That's why many students with working memory challenges are late bloomers. Although they struggle a lot during the early and middle grades, they make significant progress during adolescence and young adulthood.

Finally, working memory is also a key component of attention and *executive functioning* skills like organization, planning, implementation, and oversight of tasks. That's why working memory limitations can contribute to challenges in these areas. Often students with severe working memory challenges are diagnosed with inattentive attention deficit hyperactivity disorder (ADHD). This is especially common in dyslexic students who also show challenges with procedural learning or processing speed. Overall, about 25 to 40 percent of dyslexic students will also meet the qualifications for ADHD (and vice versa).

Processing Speed

The sixth and final key brain processing function is *processing speed.* Processing speed, as its name suggests, reflects the speed with which an individual can take in, process, and respond to information. Processing speed can be lowered by many of the characteristic dyslexic brain differences we've mentioned, such as the use of physically larger brain circuits, poor development of automaticity, increased neural noise, and higher firing thresholds for neurons.

Processing speed is one of the items usually included on IQ and other psychometric tests. Low processing speed is unlikely by itself to produce dyslexic-type challenges. However, it usually increases the impact of the other challenges discussed in this section. The aspect of reading that slow processing speed generally affects most is reading fluency (i.e., reading quickly, accurately, and automatically). Slow processing speed also contributes to challenges with attention and executive functioning, and a co-diagnosis with ADHD is very common in dyslexic students with very slow processing speeds.

Protective Factors in Dyslexia

Although the "challenge factors" we've just discussed are usually the main focus of dyslexia researchers, some have begun to look for "protective" or "resilience" factors that can limit the problems caused by such issues. Dr. Fumiko Hoeft, a professor of psychological sciences and the director of the Brain Imaging Research Center at the University of Connecticut, has studied such factors extensively. As Fumiko told us, "There are a lot of kids who have risk factors for dyslexia but who learn to read and spell better than others with similar risks. So the question we've asked is 'What are the factors underlying this resiliency?' We wrote a paper a few years back where we described the kinds of *cognitive* and *emotional* resilience factors that are important for dyslexic students to do well at various stages." The cognitive factors Fumiko and her colleague identified included:

- strong language, vocabulary, and general knowledge skills
- strong executive function and attention skills
- strong fine motor skills, which correlate with better attention and procedural learning

The social and emotional factors they identified included:

- self-awareness, proactivity, perseverance, realistic planning and goal setting
- strong sense of self-determination, hope, and control over one's choices and actions
- a growth mindset, or positive belief in the ability to learn and develop one's talents through hard work, planning, and learning from others
- strong and supportive relationships with parents (especially parents with a good understanding of their learning challenges), teachers, and one or several peers

Like the challenge factors, these protective factors can be thought of as dimensions in our dyslexia model. They are positive dimensions, like clear visibility or low wind speed on a mountain trip. While they are clearly strengths, they are not the kinds of dyslexic advantages that are the focus of this book, because although they can help dyslexic individuals read and perform other tasks better, they are not an inherent part of the dyslexic profile. We'll discuss social, emotional, and other factors important for building a positive self-image in detail in chapter 34.

Dyslexic Challenges Beyond English

Finally, let's take just a moment to consider what dyslexia looks like in different languages. Dyslexic reading difficulties have been found in every language studied. In each of these languages, dyslexic reading challenges are caused primarily by the six challenge factors we've just discussed, though their relative importance varies depending on the nature of the language.

There are two main types of printed languages. The first are *alphabetic languages,* in which symbols are used to represent each of the different sounds in their language. The second are *phonographic (or pictorial) languages,* like Chinese languages or Japanese, in which symbols depict whole words or larger units of sound, like syllables, that contain combinations of sounds.

In both kinds of languages, individuals with poor naming speed struggle to become fluent readers. Visual processing difficulties also impair reading fluency in most languages.

The impact of word sound or phonological processing difficulties varies more between languages. In general, phonological processing challenges create more problems in alphabetic than pictorial languages; and among alphabetic languages, the impact of the challenges varies according to the transparency—that is, the simplicity and consistency—of the rules that govern spelling in that language. In less transparent languages like English, where rules are complex and have many exceptions, phonological deficits often cause problems with decoding, spelling, and reading fluency. In more consistent and transparent languages, like Italian or Finnish, fluency is often a bigger problem than decoding or spelling. Even in a pictorial languages like Mandarin, phonological processing problems can still cause difficulties. Although Mandarin symbols don't represent *single* phonemes, like the letter symbols in alphabetic languages, many Mandarin symbols do represent *blends* of phonemes, and phonological deficits can impair the identification and manipulation of these sounds.

Conclusion

In the last three chapters we've looked for key features in dyslexic brains that can help to explain the remarkable duality of dyslexic strengths and challenges. We've also begun to identify the dimensional features for our model of dyslexia. In this chapter we've looked primarily at some of the challenges that form the flip-side trade-off of the dyslexic advantage. In the chapters that follow, we'll show you the advantages themselves.

Part 3

M-Strengths

Material Reasoning

7

The "M" Strengths in MIND

M-Strengths in Action: Catherine Drennan

It has been known since ancient times that form and function interact in important ways. However, this connection has never been more closely or productively explored than in recent studies of biological molecules like proteins and nucleic acids. Most people know how James Watson and Francis Crick's discovery of DNA's double helical structure helped us understand how genetic traits were inherited and how proteins were encoded and produced. Similarly, studying the shapes and forms of other biological molecules has also helped us understand their functions and allows us to modify or mimic those functions for medical or industrial uses.

Although many ingenious devices have been created to make the analysis of molecular structures easier, interpreting the data these devices produce still falls to human researchers. It is an almost bafflingly complex business, and it takes a special kind of mind to translate the two-dimensional shadow patterns created by the scattered X-rays that these devices emit into complex, dynamic three-dimensional spatial representations. In fact, it takes a mind like Catherine Drennan's.

Cathy is a professor of chemistry and biology at the Massachusetts Institute of Technology, one of the world's top research universities. Cathy's research specialty is using techniques like X-ray crystallography to investigate the structures of metalloproteins, or proteins that incorporate a metal ion into their structures. In addition to being a world leader in her highly elite field, Cathy has won many teaching awards at MIT, where her students describe her as a "brilliant scientist, engaging teacher, dedicated mentor, and fierce advocate."

By anyone's definition, Cathy is a standout performer. She is also someone who sits at the very peak of the educational pyramid, teaching the future university professors who will teach the ed-school teachers who will teach the classroom teachers who will teach the rest of us. It's easy to assume that such a teacher of teachers must have been the kind of student who succeeded spectacularly from her first day in the classroom; but if you were to assume that, you would be very, very wrong.

Cathy was the only child of two older and intellectually active parents. Early on, Cathy's parents welcomed her as a third party into their adult conversations on favorite topics such as politics, economics, and science. When Cathy began first grade, her teacher was immediately impressed with her conversational skills and wide range of knowledge. When it came time to assign students to reading groups, Cathy's teacher assumed that her spoken language abilities would translate into rapid success in reading, so she placed Cathy in a group with the students she thought would learn fastest.

Unlike the other students in this group, Cathy made little

progress, and she was soon far behind. Her teacher dropped her down to the second reading group, but Cathy fell behind there, too. She was dropped one more level. Then another. By the end of the year she was in the lowest reading group of all, and she was the very worst reader in that group.

Cathy's school gave her some tests, then called her parents in for a meeting. Cathy had dyslexia, her parents were told, and unfortunately it was so bad she would likely never graduate from high school.

Cathy was taken out of the regular class and put into special ed. Her parents arranged for extra tutors outside of school. None of the attempts at remediation seemed to help, and by the end of sixth grade, Cathy still had not learned how to read. The school told her she could not go on to middle school until she could learn to read better. So the next fall, when her peers moved on to seventh grade, Cathy returned to sixth grade for the second time.

Slowly, unexpectedly, things began to change for Cathy. Although she could still not sound out words using the techniques her teachers tried to show her, she realized through an insight of her own that she could distinguish and recognize words by their different shapes. She used things like word length, shapes of first and last letters, and places where letters spiked up or down from the line (with *b* or *p*, for example) to recognize the unique patterns that distinguished different words. Cathy already knew the definitions and sounds of a large number of words from her experiences with spoken language. Now she learned how to match these words with the "pictures" that were used as symbols to represent those words on the page.

By the end of that year Cathy was reading well enough to advance to seventh grade. She also moved out of special ed and back into the regular classroom. By the end of high school, she had worked so hard that she made up all the time she had lost and was able to graduate with her original class. She was even accepted at Vassar, a highly regarded college in New York.

In college, Cathy fell in love with chemistry and biology. She soon decided that she wanted to be a scientist. However, that would require graduate school, and one of the prerequisites for admission to graduate school is a standardized test called the Graduate Record Examination (GRE). This was a real problem, because the GRE entails a lot of reading under time-limited conditions. Although Cathy could now read with good comprehension, she still read quite slowly. She knew that unless she was given extra time to take the test, she could not do well enough to qualify for the kinds of top-tier graduate programs for which her grades and letters of recommendation would otherwise qualify her.

Cathy had to jump through many hoops to get that extra time, but her years of struggle had nurtured her persistence. She stood up for herself and was finally given the accommodation she needed. Her efforts paid off when she did well enough to earn a spot in the excellent graduate school at the University of Michigan.

Soon after arriving at Michigan, Cathy found an adviser, Professor Martha Ludwig, who "took me under her wing, mentored me, and helped me reach my full ability. I didn't have lot of confidence, and I had a really huge impostor syndrome, like I was waiting for everyone to figure out that I really didn't belong there. I always felt like there was going to be this sort of dyslexia ceiling, like . . . you got this far, but no farther—you're done. But Martha saw in me this great potential, and gave me tremendous opportunities, like an opportunity to talk about my research in an international meeting." This last opportunity led directly to Cathy's being offered a position at MIT.

In time Cathy came to believe that "it was an advantage to be dyslexic in some of the research I was doing." Her research involved visualizing tiny protein molecules and how they change shape as they perform their functions. Because there isn't any microscope capable of visualizing things so small, scientists have to use methods like X-ray crystallography, where they shoot X-rays through crystals made of the proteins to create diffraction patterns. Using complex mathematics, they can then create three-dimensional density maps that show where all the electrons are in the protein. However, those maps don't show which electrons belong to which parts of the protein. That part of the process is highly dependent upon the spatial reasoning of the scientist.

Cathy told us what's involved: "Proteins are made up of twenty amino acids, and as a researcher, you know what the electron density maps of each of those looks like. You also know the order of amino acids in your protein. And you have your electron density maps of the whole molecule. So now you have to figure out which amino acid goes in which piece of electron density.

"This is the perfect technique for me, because this is exactly how I learned to read: I was given the shape of a word, and I had to figure out what word went into that shape. I'm very good at recognizing patterns that other people don't see in these density maps and seeing things in that density that other people miss . . . because I have years of experience of looking at patterns. It's what I do. I really do believe that there are things that you see differently when you are dyslexic, which can be a huge advantage in certain areas. This strength in visualizing, this ability to see shapes and patterns, really is an advantage for me."

In reflecting on her experiences, Cathy shared these thoughts: "Learning disabled: That was a term my elementary school teachers used for me. So what is being learning disabled? When you think about that, it means you learn differently. If there's a bunch of information, you may see it in a different way. You may make connections between that information that other people miss. In those ways, it has a lot in common with what we often mean when we say *smart*."

Cathy has some advice for young people with dyslexia: "Don't listen when people want to tell you what you can't do. There is no ceiling for people with dyslexia. It doesn't exist unless you create it in your own mind."

We've shared Cathy's story with you because, as you'll soon see, Cathy is a perfect example of a dyslexic individual who excels in material reasoning, or the M-strengths in MIND.

Material Reasoning: A 3-D Advantage

> M-strengths are the 3-D spatial reasoning abilities that help us reason about the physical or material world—that is, about the shape, size, motion, position, or orientation in space of physical objects, and the ways those objects interact.

M-strengths consist of powerful *three-dimensional spatial reasoning,* and they arise from the ability to construct accurate 3-D mental models of real and imaginary physical spaces and objects and to manipulate those models at will. Spatial reasoning has often been recognized as an area of special talent for many individuals with dyslexia. However, as we'll show in the next few chapters, dyslexic individuals with prominent M-strengths typically possess outstanding abilities in some areas of spatial reasoning but not others.

While M-strengths receive little emphasis or nurturing in most school curricula, they play an essential role in many adult occupations. Designers, mechanics, engineers, surgeons, radiologists, electricians, plumbers, carpenters, builders, skilled artisans, dentists, orthodontists, architects, chemists, physicists, astronomers, drivers of all types (truck, bus, and taxi), and computer specialists (especially in areas like networking, program and systems architecture, and graphics) all rely on M-strengths for much of what they do.

In the coming chapters, we'll look in detail at the nature and advantages of M-strengths and at the key mental processes underlying them.

8

The Advantages of M-Strengths

Let's begin our examination of M-strengths by looking at two studies that compare the performance of individuals with and without dyslexia on various spatial tasks.

In the first study, British psychologist Elizabeth Attree and her colleagues compared dyslexic and nondyslexic adolescents on three different visual and spatial tasks. The first two tasks were designed to assess two-dimensional spatial skills. For the first task, the subjects were shown printed 2-D patterns, then were asked to reproduce them using colored blocks. For the second, the subjects were shown abstract line drawings for five seconds and then were asked to reproduce them from memory.

The third task was designed to assess three-dimensional spatial skills—in particular, the kind of 3-D skills needed to perform well in a real-world spatial environment. For this task, subjects were seated before a computer screen and asked to search through a virtual 3-D house to find a toy car hidden in one of the rooms. After they'd searched the four virtual rooms, the computer was turned off and they were asked to perform a task that was unrelated to their original instructions: to reconstruct the house's floor plan from memory using architectural blocks.

The dyslexic and nondyslexic groups performed very differently on the 2-D and 3-D tasks. The dyslexic individuals did slightly worse than nondyslexics on the 2-D tasks (which stressed simple snapshot visual memory and required correct directional orientation in two-dimensional space). However, they did much better on the virtual reality task, where they had to construct a seamless interconnected world from the views they'd absorbed during their explorations.

Let's pause for a moment to consider two aspects of these results that will come up repeatedly in our discussions of the MIND strengths. First, notice the contrast between the weaknesses shown by the dyslexic subjects in the 2-D tasks, which required fine-detail processing and snapshot memory, and their strengths in the 3-D task, which required them to create a complex holistic interconnected mental model of the virtual house. This is yet another example of the dyslexic trade-off between fine-detail and big-picture processing that we've noted already. Second, notice how difficult the dyslexic subjects found it to remember abstract shapes or figures, which had no distinct purpose, character, or meaning, in contrast to their ability to remember their experience of navigating through the more real-world virtual space. This difference between rote memory of abstract information and memory for meaningful associations and experiences is deeply characteristic of dyslexic minds, and will be discussed in more detail in later chapters.

This marked divergence in ability between different spatial tasks has important implications not only for how we think about dyslexic strengths and about the concept of spatial ability in general, but also for how we assess spatial abilities on tests. Many of the visual-spatial tasks commonly used in IQ tests and other cognitive batteries (such as block design and visual memory tasks) assess 2-D spatial abilities and fail to measure the kinds of real-world 3-D spatial strengths that individuals with dyslexia possess.

A second study, performed by psychologist Catya von Károlyi and her colleagues, also supports the existence of a dyslexic 3-D/2-D trade-off. Károlyi compared the abilities of dyslexic and nondyslexic high school students on two visual-spatial tasks. For the first task, subjects were asked to find an identical match for a complex 2-D Celtic knot pattern from a group of four closely related patterns. This task required great accuracy in processing visual fine details. For the second task, subjects were shown a series of pictures, each of which displayed four similar shapes, and were asked to determine which of the four shapes in each picture represented an impossible figure that could not exist in 3-D space. Success on this latter task required the ability to perceive how different parts of a figure related to each other in forming a larger whole—thus, big-picture or global processing.

The results on these two tasks perfectly mirrored those in Dr. Attree's study. On the 3-D impossible figures task, Dr. Károlyi found that individuals with dyslexia answered significantly more quickly than, and just as accurately as, the nondyslexic group. In contrast, on the 2-D Celtic knot task, individuals with dyslexia were significantly less accurate than their nondyslexic peers.

At first glance, this skill in detecting impossible figures might seem to have little real-world value, but we found an excellent example of its practical significance when speaking with a highly successful building contractor. This contractor, who is himself dyslexic, told us he prefers hiring dyslexic workers for his building crews because they excel at spotting flaws in blueprints that create impossible figures just like those in Dr. Károlyi's study.

This ability to mentally simulate 3-D spatial information is actually extremely valuable in many real-world occupations, and we can get further insight into the extent of this value when we examine how individuals with dyslexia put these skills to use at various stages throughout their lives.

The Real-World Worth of M-Strengths

Early in life, many dyslexic children seem naturally drawn to engage in highly spatial tasks. In a survey of children from our clinical practice (ages seven to fifteen), we found that children with dyslexia engaged in building projects—everything from LEGOs and K'NEX to small models to massive outdoor landscaping and construction projects—at nearly twice the rate of their nondyslexic peers. Even when these children engaged in 2-D art projects like drawing, their art tended to have a more multidimensional and dynamic quality, featuring elements like foreshortening and perspective, moving figures, arrows indicating action or process, and schematic elements like cutaway sections or multi-angle or multi-perspective blueprints. Dr. Jean Symmes, who was a research psychologist at the National Institute of Child Health and Human Development, similarly documented an unusually high interest and ability in building or visual classification tasks among the children with dyslexia she studied.

While it's sometimes claimed that children with dyslexia gravitate toward high M-strength activities (and later occupations) because their reading and writing challenges make other activities too difficult, the preceding observations suggest that for most spatially talented dyslexic individuals, spatial interests and abilities actually predate their reading challenges. Former Harvard neurologist Dr. Norman Geschwind—one of the most esteemed figures in the history of dyslexia research—noted that in his experience many dyslexic children display a passion and skill for spatial activities (like drawing, doing mechanical puzzles, or building models) well before they begin to struggle with reading.

As children grow up, a more direct link between dyslexia and high M-strength occupations becomes visible. Several studies have demonstrated that individuals with dyslexia are significantly overrepresented in training programs for highly spatial professions like art, design, and engineering. In the United Kingdom, where only about 4 percent of the population is considered severely dyslexic and another 6 percent moderately dyslexic, a study at the Royal College of Art found that fully 10 percent of its students showed severe dyslexic findings, and 25 percent at least moderate findings—*more than double the rates in the general population*. At the Central Saint Martins College of Art and Design in London, psychologist Dr. Beverley Steffert found that more than 30 percent of the 360 students she tested showed evidence of dyslexia-related difficulties with either reading, spelling, or written syntax. Another student survey at the Harper Adams University in England showed that 26 percent of the first-year engineering students were dyslexic. In Sweden, researchers Ulrika Wolff and Ingvar Lundberg compared the incidence of dyslexia in university students majoring in fine arts and photography with a control group of students studying economics and commercial law. They found that the incidence of dyslexia was nearly three times greater among art students than among either the control students or the general Swedish population.

While formal studies on the incidence of dyslexia among fully qualified professionals in these fields are lacking, there is no shortage of occupational lore. In his book *Thinking Like Einstein*, author Thomas G. West recounts his conversation with dyslexic computer graphic artist Valerie Delahaye, who specializes in creating computer graphic simulations for movies. Delahaye told him that at least half the graphic artists she's worked with on major projects like *Titanic* and *The Fifth Element* were also dyslexic. West quotes MIT Media Lab founder and dyslexic Nicholas Negroponte, who stated that dyslexia is so common at MIT that it's known locally as the "MIT disease." UCLA professor Maryanne Wolf has written that spelling difficulties are so widespread at the architectural firm where her brother-in-law works that they've instituted a rule that all architects must have their outgoing letters spell-checked—twice. And author Lesley Jackson wrote in the design industry trade journal *Icon Magazine*: "Having met so many dyslexic designers over the years, I've become convinced there must be some kind of link between the underlying processes of design creativity and the workings of the dyslexic mind."

Many dyslexia experts have also gone on record with their own observations regarding the links between spatial ability and dyslexia. Dr. Norman Geschwind wrote: "It has become increasingly clear in recent years that individuals with dyslexia themselves are frequently endowed with high talents in many areas. . . . [A]n increasing number of studies . . . have pointed out that many individuals with dyslexia have superior talents in certain areas of non-verbal skill, such as art, architecture, [and] engineering." Eminent British neurologist Macdonald Critchley, who personally examined more than 1,300 patients with dyslexia, stated that "a great many" of these patients had shown special talents in spatial, mechanical, artistic, and manual pursuits, and that they frequently pursued occupations that made use of these abilities. We could easily cite many more such observations.

The Cognitive Basis of M-Strengths

To reason effectively about movement, position, or shapes in 3-D space, the brain must be able to form an accurate mental model of the 3-D world. In recent years, researchers led by 2014 Nobel Prize winners John O'Keefe at University College London and May-Britt and Edvard Moser at the Norwegian University of Science and Technology have made tremendous progress in explaining how this 3-D model is constructed. As we'll see throughout this book, their work has also helped set in motion a fascinating series of discoveries that are relevant not only for spatial reasoning but for all the MIND strengths.

What O'Keefe and the Mosers discovered were sets of specialized cells in the brain's hippocampus* and surrounding structures that interact to create a virtual 3-D model in the brain. The Mosers identified cells they named *grid cells* that function to create the coordinate lines in the brain's 3-D spatial model. If it helps, you can picture these intersecting lines as the bars of an infinite jungle gym. This spatial matrix allows us to plot information about where objects are in space—much like a 3-D GPS navigation system. O'Keefe discovered *place cells* that fire only at specific locations in 3-D space, providing information about objects and events that are (or have been) located at those positions in the 3-D grid. Together, grid cells, place cells, and several other types of cells interact in ways that allow us to monitor not just our own positions in space but also the positions, shapes, and movements of other objects in our real-time environment. They also create a

---

* The hippocampus is a complex structure at the base of the brain whose two seahorse-shaped lobes play many key roles in memory formation and learning.

kind of mental workshop where we can simulate the spatial forms, positions, and movement of objects that we don't currently see; imagine the spatial forms of entirely new objects; plan and envision our own future movements through space; or perform any other kind of 3-D spatial reasoning process.

This spatial information that we process in our minds can be presented or displayed to our conscious awareness as various forms of *spatial imagery*. The most obvious form of spatial imagery is visual, where a person sees clear, vivid pictures, shapes, movements, and colors. An excellent example of a dyslexic individual with impressive M-strengths and a remarkably clear and lifelike visual display of spatial imagery is Canadian entrepreneur Glenn Bailey. After academic problems caused him to drop out of high school, Glenn became a highly successful businessman. One of his many successful ventures has been the development and construction of residential real estate. Glenn described for us how his ability to generate and voluntarily manipulate vivid lifelike 3-D visual imagery often helps him in this business. "When I see a property, I can instantly construct a new house on it. I can see exactly how that house is going to look, and I can walk through every room in that house and out into the garden and everywhere. I can turn those thoughts into reality. And that's how my development company was created for high-end houses. Even right now, sitting here, I can do a detailed walk-through in my mind of every house and property we've ever built."

Although stories like Glenn's might cause you to assume that the ability to see clear and lifelike pictures in the mind is essential for spatial reasoning, the experience of MX shows clearly that this assumption is incorrect. MX was a retired building surveyor living in Scotland who'd always enjoyed a remarkably vivid and lifelike visual imagery system, or mind's eye. Unfortunately, four days after undergoing a cardiac procedure, MX awoke to discover that though his vision was normal, when he closed his eyes, he could no longer voluntarily call to mind any visual images at all—a condition known as *aphantasia*.

MX was tested using a whole series of spatial reasoning and visual memory tasks. As a control, MX's doctors had a group of high-visualizing architects perform the same tasks. Surprisingly, they found that although MX could no longer create any mental visual images while performing these tasks, he scored just as well on them as the architects did. As he performed the tasks, MX's brain was also scanned with fMRI technology. In contrast to the architects, who heavily activated the visual centers of their brains while solving these tasks, MX used none of his brain's visual processing regions.

These studies suggested that while MX had lost his ability to *perceive visual images* when engaging in spatial reasoning, he could still *access and manipulate spatial information* in his spatial modeling system and apply it to material reasoning tasks with no detectable loss of skill. In other words, MX had gone quite literally overnight from having remarkably vivid visual imagery to having none at all, without any apparent loss in his spatial reasoning abilities. This is a dramatic demonstration of the difference between spatial reasoning ability and visual imagery.

Spatial imagery can actually be perceived in many ways besides clear, lifelike visual forms. As long as the hippocampus and its networked cortical structures can create their spatial grid from information gathered through the senses, it seems relatively less important what form of imagery the individual uses to read or access this information. Think, for example, of a blind person who recalls the contours of a friend's face: this spatial information is recalled in a nonvisual form, as a form of tactile or muscular (somatosensory) imagery, yet it can be every bit as accurate and detailed as visual imagery.

We can also demonstrate the variety of useful spatial imagery styles by examining what other individuals with dyslexia and impressive M-strengths have said about their own forms of spatial imagery. Let's start with the legendary physicist Albert Einstein.

In addition to having remarkable M-strengths, Einstein showed many dyslexia-related challenges, such as late talking, difficulty learning to read, a poor rote memory for math facts, and a lifelong difficulty with spelling. Einstein described his own spatial imagery in the following way: "The words of the language, as they are written or spoken, do not seem to play any role in my mechanism of thought. The psychical entities which seem to serve as elements in thought are certain signs and more or less clear images which can be 'voluntarily' reproduced and combined."

This kind of abstract imagery is especially common among spatially talented physicists and mathematicians, for whom the flexibility of such imagery seems to be particularly valuable. Dyslexic mathematician Kalvis Jansons, another professor at University College London, has written: "To me, abstract pictures and diagrams feel more important than words. . . . Many of my original mathematical ideas began with some form of visualization."

Jansons has also described experiencing spatial imagery in a completely nonvisual form—as feelings of movement, force, texture, shape, or other kinds of tactile or motor images: "It would be a mistake to believe . . . that non-verbal [spatial] reasoning has to involve pictures. For example, three-dimensional space can be equally well represented in what I often think of as a tactile world." Jansons has employed this "tactile" spatial imagery in his professional work by using knots to study important principles of probability.

Dr. Matthew Schneps of the Harvard-Smithsonian Center for

Astrophysics shared with us a related form of spatial imagery. Matt is an astrophysicist, an award-winning documentary filmmaker, and an individual with dyslexia. Matt described his spatial imagery to us as consisting of a feeling of movement or process—rather like a machine at work. When pursuing an idea or hypothesis, Matt sometimes feels like he's activating a lever in a machine that he imagines in mental space, in order to turn a series of gears and observe how the spatial map changes as he tests various configurations.

Dyslexic attorney David Schoenbrod described to us still another type of nonvisual spatial imagery. David is Trustee Professor of Law at New York Law School, a pioneer in the field of environmental law, and a key litigator in some of the most important environmental cases of the last fifty years, including the landmark lawsuits that led to the removal of lead from gasoline. He is also a talented sculptor, architect, landscape designer, and builder. David described his spatial imagery to us as a strong sense of spatial position unaccompanied by clear visual images: "In recalling autobiographical stories I recall spatial arrangements in some detail—like the layout of the room, the arrangement of the furniture, where the other people and I were, and the ordination to the points of the compass—but this recollection is neither lifelike nor schematic, but rather in gray scale, and almost vanishingly faint. In general, I know the shape of things, but I don't really see them. It strikes me that the fact that I see form more than color is why I have been more attracted to sculpting than painting."

We've described these different forms of spatial imagery in some detail because we too often meet educators and individuals with dyslexia who believe that lifelike visual imagery is the key to spatial reasoning. As a result, they often overlook the value of other forms of spatial imagery. In truth, it doesn't seem to matter much whether your mental imagery consists of lifelike visual images or whether it is positional or movement or touch related. So long as you can use this imagery to understand spatial relationships, you can use it to make important comparisons and predictions, or to combine, change, or manipulate spatial data in various ways. However, as we'll discuss in the next chapter, M-strengths often come with several trade-offs.

9

Trade-offs with M-Strengths

Each of the MIND strengths comes with its own set of trade-offs, and M-strengths are no exception. We've mentioned one trade-off already: a relative weakness in certain two-dimensional processing skills. While this weakness has few consequences for most day-to-day functions, there's one area where it can create important problems: symbol reversals while reading or writing.

Struggles with Symbols

We often encounter two opposite and equally mistaken beliefs about symbol reversals and dyslexia. The first is that all young children who flip symbols are dyslexic. The second is that a tendency to reverse symbols is *never* associated with dyslexia. To sort out the truth about this topic, we must examine how human spatial skills develop.

No child is born with the ability to identify the two-dimensional orientation of printed symbols—or of anything else, for that matter. The ability to distinguish an object from its mirror image is actually an acquired skill that must be learned through experience and practice.

Over the last decade, researchers have found that the newborn human brain forms two mirror-image views of everything it sees: one in the left hemisphere and the other in the right. Usually this duplicate imagery is helpful because it allows us to recognize objects from multiple perspectives, so that a toddler who's been warned about a dog while looking at its left-side profile can recognize that same dog when looking at its right-side profile.

Unfortunately, when trying to recognize the orientation of printed symbols—or any other item with a natural mirror, like a shoe or glove—this ability to generate mirror images becomes a burden. Before a child can reliably distinguish an image from its mirror, he or she must *learn to suppress* the generation of its mirror image.

Some children have an especially hard time learning to suppress this mirroring function. When they first learn to write, many children will reverse not only symbols that have true mirrors (like *p/q* or *b/d*), but essentially all letters or numbers. For most children, these mistakes begin to diminish after only a few repetitions. However, until the age of eight, as many as a third of children continue to make occasional mirror-image substitutions when reading or writing. If such mistakes are only occasional and the child has no real difficulty with reading and spelling, these errors are neither important nor necessarily a sign of dyslexia.

However, for some truly dyslexic children—in our experience, perhaps as many as one in four—letter reversals can be a much more persistent and important problem. These children may reverse whole words or even whole sentences, and at the single symbol level they may reverse not only horizontal mirrors like *b/d* or *p/q*, but also vertical mirrors like *b/p*, *b/q*, *d/p*, *d/q*, or *6/9*. They may make so many reversals when reading that it worsens their comprehension.

Published studies have shown that many younger dyslexic children have more difficulty rapidly determining letter orientation than their nondyslexic peers, though this difficulty declines with age. In our experience, persistent reversals—not just for letters and numbers, but even for drawings and other visual figures—are most often a problem for those who are especially gifted with M-strengths. Leonardo da Vinci provides us with an extreme example of this phenomenon. His lifelong dyslexic difficulties in reading, word usage, syntax, and spelling were combined with phenomenal M-strengths. While many people are aware that Leonardo wrote his journals in mirror-image script, few know that he also drew many of his sketches and landscapes in mirror image.

We've worked with many dyslexic individuals who've continued to reverse symbols when reading—or more commonly writing—well into their college years and beyond. Most of these individuals experience sporadic errors, but we met one student who unintentionally lapsed into writing full paragraphs in mirror image whenever she grew tired. Probably not coincidentally, she obtained her graduate degree in architectural history.

One reason that spatially talented individuals with dyslexia may be especially susceptible to reversals is that their brains are just so good at rotating spatial images. Listen to dyslexic designer Sebastian Bergne: "If I'm designing an object, I know the exact shape in 3-D. I can walk around it in my head before drawing it. I can also imagine a different solution to the same problem." While this image flexibility may be useful when you're trying to design a chair or a teapot, it's less useful when you're trying to read or write symbols on a 2-D surface. Dyslexic biochemist Dr. Roy Daniels was one of the youngest members ever elected to the prestigious National Academy of Sciences, but even as an adult, he still confuses mirrored letter pairs like *b/d* and *p/q* both when reading and when writing. To compensate, he does all his handwriting in capital letters, "to help me tell the difference between letters like *b* and *d*." Dr. Daniels is far from unique in this regard.

It's likely that difficulties with procedural learning, which we discussed in chapters 5 and 6, contribute to these persistent reversals because the ability to turn off the symmetrical image generator is itself a kind of procedure that must be learned through practice. As a result, it will likely be mastered more slowly by individuals with dyslexia who show procedural learning challenges.

Ease of Language Output

A second trade-off that we often see in individuals with dyslexia and prominent M-strengths is difficulty with language output. Parents and teachers are often puzzled to find that their bright dyslexic students struggle to answer apparently "simple" questions—especially in writing, but also sometimes in speech. This difficulty can be particularly intense when the questions are open-ended and students are given a great deal of latitude in how they respond. Difficulty answering questions of this kind is one of the most common reasons why older dyslexic students are brought to our clinic. We've found that this difficulty is often particularly bothersome for dyslexic individuals with high or even gifted-level verbal IQs, because so often the ideas these students are trying to express are complex.

The research literature suggests several possible reasons why dyslexic individuals with impressive M-strengths may be especially vulnerable to difficulties putting their thoughts into words. First, some of the brain variations associated with dyslexia may enhance spatial abilities at the direct expense of verbal skills. Psychologists George Hynd and Jeffrey Gilger have described one such variation. In this structural variation, brain regions that are normally used to process word sounds and other language functions are essentially "borrowed" and connected instead to brain centers that process spatial information. Drs. Hynd and Gilger first identified this brain variation in a large family with many members who showed both dyslexia and high spatial abilities. They then identified this same variation in the brain of Albert Einstein, who, as we've mentioned, displayed a similar combination of spatial talent and dyslexia-related language challenges.

Einstein's comments on his own difficulties putting his ideas into words provide useful insight into the challenges many of our high M-strength dyslexic individuals experience. Although Einstein eventually became a talented writer, he once complained that thinking in words was not natural for him, and that his usual mode of thinking was nonverbal. To communicate verbally he needed first to translate his almost entirely nonverbal thoughts into words. Einstein described the process this way: "[C]ombinatory play [with nonverbal symbols] seems to be the essential feature in productive thought—before there is any connection with logical construction in words or other kinds of signs which can be communicated to others. . . . Conventional words or other signs have to be *sought for laboriously* [italics added] only in a secondary stage."

We've found that many individuals with dyslexia—and especially those with prominent M-strengths—identify closely with Einstein's descriptions both of his primarily nonverbal thinking style and of his difficulties in translating his thoughts into words. While translating nonverbal thoughts into words can be difficult at any stage of life, it is often especially difficult for children and adolescents, whose working memory capacities are still far from fully developed. This is likely one reason why children from families with a high degree of spatial and nonverbal attainment are often slower than other children to begin speaking.

It's important that those who work with high M-strength individuals be sensitive to this challenge. There's a long and quite shameful tradition among certain psychologists and educators of treating nonverbal reasoning as if it were at best a poor cousin of verbal reasoning and at worst a kind of oxymoron—like "civil war" or "act naturally."

In fact, nonverbal reasoning is a real, scientifically demonstrable, and often key component of creative insights of all kinds, and it deserves to be taken seriously in all its forms. While students with dyslexia should try their best to express their thoughts in words, it's also critical that parents, teachers, and later employers learn to recognize that some valid forms of reasoning may be difficult to put into words and may be better expressed as drawings, diagrams, or other forms of nonverbal representation.

Besides this fairly direct trade-off between spatial and verbal ability, studies have also demonstrated a more indirect way that strong spatial and visual imagery skill can hinder verbal functions. Dr. Alison Bacon and her colleagues at the University of Plymouth in England asked dyslexic and nondyslexic college students to supply a valid conclusion to a series of syllogisms for which they'd been given major and minor premises. For example, if they were given the premises "All dogs are mammals" and "Some dogs have fleas," they were asked to provide a conclusion, such as "Some mammals have fleas."

The researchers found that the dyslexic students reasoned just as well as their nondyslexic peers when they were given premises that provoked little imagery (e.g., "All *a* are *b*, no *b* are *c*, how many *a* are *c*?"), or when the visual imagery contributed directly to the solution of the syllogism (e.g., "Some *shapes* are *circles*, all *circles* are *red*, how many *shapes* are *red*?"). However, when the syllogisms contained terms that provoked strong visual images that were *unrelated to the reasoning process* (e.g., "Some snowboarders are jugglers, all horsewomen are snowboarders, how many horsewomen are jugglers?"), the dyslexic students performed significantly *worse* than the nondyslexics. The authors concluded that their vivid mental imagery was swamping their working memory and hindering their verbal reasoning.

This potentially distracting role of visual imagery has important implications for how we teach dyslexic students with M-strengths. Think, for example, of how needlessly burdened a student with strong imagery abilities will be by visually elaborate story problems in math. Many teachers have been taught that using imagery helps children with strong spatial and visual skills, but this is true only if the imagery is directly useful for solving the problem. Irrelevant imagery is distracting and worsens performance.

A final point to remember about language development in individuals with dyslexia—and especially in those with prominent M-strengths—is that their language is simply developing along a different pathway than the one followed by their nondyslexic peers. The brain systems that help translate nonverbal ideas into words—including working memory—are some of the latest-developing parts of the brain. For many children and adolescents with dyslexia, difficulty putting complex ideas into words is a normal feature of development and one that diminishes with maturity. That's why their progress must be judged by its own standards, rather than by standards that apply to the nondyslexic population. Focusing too much on their challenges can make us overlook their special strengths, as we observed with one very special child, Max.

10

Growing Up with M-Strengths

M-Strengths in Action: Max

As an infant, Max was late to start talking, and when he finally began to speak, it was in a language all his own: *ma* was water, *dung gung* was vacuum cleaner, and *wow wow* was pacifier. When he started preschool at age three and a half, Max's mother recalled, he had difficulty "catching on to things that the other kids seemed to simply absorb. He never learned the songs or rhymes, couldn't remember the names of the other kids, and could rarely retell what happened during the day." In first grade he went to a Montessori school, but "he didn't 'discover' academic knowledge and skills on his own. He needed to be explicitly taught."

Through the end of first grade Max made little progress in reading, math, or writing. He seemed to have a hard time staying focused. He also struggled to retrieve words and information from memory. His kindergarten and first-grade teachers found that he seemed to learn much better one-on-one than in a large class, so Max's mother decided to homeschool him for second grade.

Max slowly began to learn. Although he still required frequent repetitions and refocusing, by the end of second grade he was reading—slowly—and his writing began to take off (though it largely left his spelling behind). The following is a response he wrote at age nine to the question "Tell me about going to the [Seattle] Science Center":

> we went a long wae and thin we wint in sid. And we qplab [played] with the ecsuvatr [excavator] and thin we trid too pla with the tic tac toe mushen [machine] and thin we wint too the bug thing and thin we wint too the binusho [dinosaur] thing and thin we wint toe the ecsuvatr and thin we left.

Outside of school, Max found plenty to occupy his time. When he was a toddler, he became fascinated with wires and circuits, and as he grew older, he developed a sophisticated interest in electronics. He especially liked experimenting with small-scale power generation, using sources like solar, wind, and water to generate electricity.

Max also developed a deep interest in nature, and he loved to spend time in the woods surrounding his home near Seattle. Max laid out a nature trail on his family's property, but as many Northwesterners have learned, he soon found that *walking* plus *woods* equals *wet feet.* So Max began to install an elaborate series of drains to remove the standing water that collected across his trail. He also built bridges over the spots he couldn't drain. Max's drainage project was remarkable for a child not yet ten years old; and perhaps not surprisingly, it did draw remarks—from the psychologist Max's mother took him to see in fourth grade to be evaluated for his difficulties with schoolwork.

The psychologist diagnosed Max with ADHD. This seemed plausible given Max's problems with auditory-verbal working memory, distractibility, and lack of focus for schoolwork. However, the psychologist was also concerned by Max's intense interest in electronics and drainage, his focus on solitary pursuits, and by his difficulties talking with—and like—other children. So in addition to ADHD, the psychologist diagnosed Max with Asperger's syndrome, an autism spectrum disorder. As one of his recommendations, the psychologist suggested Max take social skills classes.

Although Max's mother questioned the Asperger's diagnosis, she agreed that Max needed to improve his social skills, so she took him to see a speech-language pathologist (SLP). Fortunately, the SLP understood that social skills consist largely of complex rules for behavior that have been learned and practiced until they've become habits. In fact, mastering social skills requires many of the same procedural learning skills we've already discussed, and is also hindered by slow processing speed and executive functioning issues. Under the SLP's supervision, Max learned about social skills in a clear and explicit manner, and he practiced them during structured interactions with another child until these skills became automatic.

Between ages seven and a half and ten, Max continued to improve both socially and academically. His reading vocabulary shot up from the 35th to the 98th percentile, and his math calculation rose from the 45th to the 99.9th. However, his IQ testing still revealed the lower working memory and processing speed scores that we typically find in our "young engineers."

At that point, Max was making great strides in most areas,

though his reading comprehension and fluency lagged behind his conceptual abilities, and his writing remained slow. He also made frequent errors in his writing with spelling, conventions, sentence structure (syntax), and organization.

We met Max shortly before his eleventh birthday. While the "numbers" from our testing mostly confirmed what others had found, our interpretation was somewhat different. We identified his challenges with reading and writing fluency, spelling, syntax, rote and working memory, focused attention for auditory-verbal material, processing speed, sequencing, and organization, but we also found many of the wonderful strengths that individuals with dyslexia often show. Max showed tremendous spatial and nonverbal reasoning powers, his understanding of math concepts was amazing, and his ability to interact knowledgeably on a wide range of complex scientific subjects was extremely impressive. We were also struck by the intellectual "flavor" Max displayed as he approached his work. He showed a charming naïveté and inventiveness on many tasks that children his age usually dash through without even giving a thought, so although his work was slower, it was often more creative. Max also made many interesting observations that showed how his mind was reaching out to probe the connections between ideas.

We also discovered several important details about Max's family. Max's father has a PhD in chemistry, and his mother has a degree in biochemistry—both high M-strength fields. Max's mother also has dyslexia-related processing traits and talents. Although she struggled with reading and writing as a child, she now works as a medical writer, because she has a talent for explaining complex ideas in simple terms. Both of Max's maternal grandparents were also scientists, and Max's great-grandmother has often remarked how much Max reminds her of his grandfather when he was a boy. Perhaps family resemblance provides a better explanation for Max's intense interest in drainage and erosion than autism, since his grandfather spent his career as a professor of geophysics at UCLA.

Ultimately, we made several suggestions to help Max in areas where he still struggled, but we also explained that in the most important respects Max was right on target for *his* development—that is, for the kind of late-blooming growth and maturation that children display when they combine dyslexic processing, outstanding M-strengths, and procedural learning challenges.

M-Strengths and Development

We've shared Max's story because we want you to see what dyslexic children with impressive M-strengths look like while they're still developing—before their mature talents are fully apparent. While it's enormously helpful to look at the childhoods of successful dyslexic adults, sometimes the perspective provided by hindsight can make their successes seem almost inevitable—as if their challenges weren't really so severe, and they were never truly at risk of failure. Somehow these stories can lack the power to convince us that the slow, awkward, inarticulate, inattentive, and dyslexic early elementary child before us might actually have a chance to become a skilled engineer, architect, designer, mechanic, inventor, surgeon, or builder. Yet this is often unquestionably true. In the years since we wrote about Max in the first edition of this book, he has gone on to earn a computer science degree and now works as a software engineer at a start-up company.

We sometimes meet skeptics who seem to derive a strange sort of pleasure from pointing out that not every child will become an Albert Einstein or an Isaac Newton. This is true, but even

Einstein and Newton didn't *look* like what we think of as Einstein and Newton in second grade: Einstein was remembered as a slow, uncooperative child with a nasty temper who repeated everything he said (echolalia), while Newton was remembered as a simpleton whose only apparent use was to make small wooden toys for his sisters and schoolfellows.

Even though evidence of classroom success may appear thin for many high M-strength children with dyslexia, they often display their creative potential quite clearly outside of the classroom in their desires to build, experiment, draw, and create. Recall compact disc inventor James Russell building his remote-control boat at age six, or Max building his nature trail and experimenting with solar power. These activities should be taken much more seriously, because for a dyslexic child with substantial M-strengths, a toy is never just a toy nor a drawing just a doodle. These activities provide a window into their future, and failure to regard them seriously does these children a grave disservice. One amazing young child with dyslexia whom we saw in our clinic built a K'NEX structure so elaborate that it won second prize in a nationwide competition; yet when he brought it to school and asked his teacher if he could show it to the class, he was told, "We don't have time for that. We have important work to do." Another was scolded by his teacher for doodling: "If you spend all day drawing buildings on your papers, you'll never get anywhere." Ironically, this child's father is a successful architect who makes his living in precisely that way.

When he was young, pioneering neurosurgeon Dr. Fred Epstein showed a sustained interest only in mechanical activities, like building elaborate model airplanes. Because of his dyslexia, Epstein barely made it through college and was initially rejected by all twelve of the medical schools to which he applied. However, as an adult, Epstein developed many new and highly innovative surgical techniques for treating previously inoperable spinal cord tumors—techniques that saved literally thousands of children's lives. It's important to realize that when Epstein was devising these techniques, he wasn't using skills he'd picked up in some classroom. He was using skills he'd developed at his workbench in the garage, building model airplanes.

To identify our next generation of talented engineers, inventors, and physicists, we shouldn't be using pencil-and-paper talent searches or seeing who's fastest at the "mad math minutes." We should be searching for spatial prodigies in LEGO stores and hobby shops—just like athletic scouts hang around ball fields. Many of our next generation's great spatial reasoners are currently struggling in school while their talents are going unrecognized, and we owe it to them to pay closer attention to the ways that they typically develop.

11

Key Points About M-Strengths

Material reasoning, or the ability to reason about the 3-D physical characteristics of objects and the material universe, represents one of the most common and important talent sets found in individuals with dyslexic processing styles. Key points to remember about M-strengths are:

- The ultimate purpose of M-strengths is to create a continuous interconnected 3-D mental model of the world, rather than focus on fine details or 2-D features.

- The spatial imagery perceived by individuals with M-strengths may take many forms, from clear visual imagery to nonvisual images involving forces, shapes, textures, or movements.

- M-strengths often bring trade-offs like symbol reversals and subtle language challenges.

- Individuals with dyslexia in general—and those with prominent M-strengths in particular—show a late-blooming pattern of development, and their developmental progress should be judged on its own terms, rather than by standards created to judge nondyslexics.

- Individuals with dyslexia who show prominent M-strengths often show signs of impressive creativity outside the classroom.

M-Strengths in Action: Jørn Utzohn

We'd like to close this section by telling you about one more person with truly remarkable M-strengths.

His name was Jørn. For those of you who did not grow up in Denmark like he did, his name is pronounced roughly like *yearn*.

Jørn always admired his father, who was a naval architect and engineer. His father designed and built boats, and he loved to sail them. Jørn shared his father's love of the sea.

After school Jørn often went down to the shipyard where his father worked. He would sit near his father and make sketches or models of the boats that his father was building. As Jørn grew older, he showed great skill as a draftsman, and his father even let him help draw some of the final plans for the boats.

Jørn decided when he grew up, he would become either an officer in the Royal Navy or a naval engineer who built boats like his father. There was only one problem. Jørn was dyslexic, and unlike his older brother, who excelled in school, Jørn struggled with reading, spelling, and doing calculations in math. No matter how hard he worked, he was never able to get good grades. Jørn graduated from high school near the bottom of his class. As a result, neither the Royal Danish Naval Academy nor any school of engineering would accept him.

Jørn had to find another path for his future, so he decided to rely on one thing he did really well: drawing. Jørn had developed excellent artistic skills both by working with his father and by drawing alongside two distinguished artists whom he often watched working by the harbor in his hometown. So he decided to study at the Royal Danish Academy of Fine Arts.

The academy's main focus was on fine arts like painting and sculpture, but it also taught design, including architecture. Jørn soon decided that architecture would be his focus. He loved to build things. He also loved nature and the natural forms he found there. He loved people, too, and he wanted to build things to make their lives better. Jørn believed that through architecture he could express the beauty and forms of nature, bringing light and natural beauty into the lives of people in ways that would make them happier and maybe even healthier.

After graduating from the Royal Academy, Jørn spent several years working for other architects. He also traveled around the world to examine the best in architecture from many other cultures and traditions. He was especially inspired by the Mayan pyramids in Central America, the sand towers in Morocco, the pagodas of China and Japan, and the works of Frank Lloyd Wright in the United States.

In 1957, Jørn took everything he had learned through his studies, work, and travels and created a series of sketches for a design competition. The people of Sydney, Australia's largest city, had decided that they wanted to build an opera house, and they wanted this building to be truly special and unlike any other building in the world. So they invited every architect who wished to compete for the honor of designing their opera house to submit plans.

Many of the world's greatest and most famous architects responded, but in the end, the design that was chosen was Jørn's.

Its mysterious beauty captured the hearts and imaginations of the competition's judges. Many said that it looked like a part of nature rather than something a human mind had designed. Some thought its curved shells looked like billowing sails. Others said it reminded them of Australia's exotic wildlife. Jørn said that he drew his inspiration in part from the structures and shapes he observed when peeling an orange.

Jørn's design for the Sydney Opera House now enchants people from around the world. It has become the most recognizable symbol of Sydney, and maybe of Australia itself. It is also one of the most beloved and widely known buildings in the world. The United Nations has even made it a World Heritage Site because of its "outstanding universal value" to humanity. It stands on their list with such timeless treasures as the Taj Mahal, Machu Picchu, the Great Wall of China, and the Roman Colosseum.

Jørn Utzon's greatest strength was his spatial imagination. His ability to translate the shapes and forms of nature into human structures has made him one of the most admired architects of our time.

12

What's New with M-Strengths

This is the first of four chapters where we'll update you on what's new and exciting with each of the four MIND strengths. In this chapter we'll focus on M-strengths, and we'll begin by reviewing the results of a study we conducted looking at how dyslexic and nondyslexic adults rate their own M-strengths.

What Dyslexic Adults Told Us About Their M-Strengths

In our research, we asked dyslexic and nondyslexic adults to rate how strongly they agreed or disagreed with thirteen statements relating to their perception of their own M-strengths. For each question, study participants were presented with a statement about a particular ability, then asked to choose which of five options best expressed how strongly they agreed that the statement accurately represented their own strengths. The five options were *strongly agree, agree, somewhat agree, disagree,* or *strongly disagree.*

As an example, here is the first statement relating to M-strengths:

I am very good at forming 3-D spatial images in my mind.

In the figure below you can see how differently the dyslexic and nondyslexic adult participants responded.

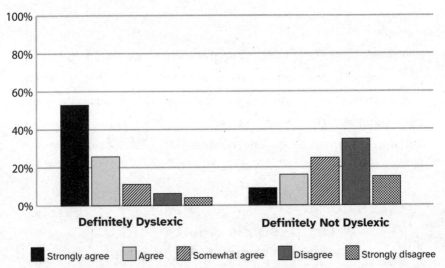

M1: I am very good at forming 3D spatial images in my mind.

The five bars on the left side of the figure show the answers of our dyslexic respondents. Almost 53 percent expressed strong agreement, while fully 79 percent either strongly agreed or agreed.

The bars on the right side of the figure show the answers of our nondyslexic respondents. Only 9 percent of nondyslexic respondents strongly agreed—a mere sixth of the rate of our dyslexic respondents—and only 25 percent agreed or strongly agreed—less than a third of the rate of the dyslexic participants.

Zoom back a bit and notice how the patterns of the bars describing the dyslexic and nondyslexic responses form essentially a mirror image. We observed this mirror-image pattern in the responses to most of the sixty-four questions we included in the final MIND strengths survey.

Unfortunately, there isn't space in this book to show you all sixty-four graphs depicting the answers to our survey. For those interested, all these charts can be found at our website dyslexicadvantage.org/mind-charts/. Here we'll simply list the questions that the dyslexic adults in our survey group agreed with significantly more strongly than did the nondyslexic ones. For those who would like to see the numerical data for all of the questions, we have included a table for each MIND strength in the Appendix.

These are the thirteen questions that our dyslexic respondents agreed with significantly more strongly than did our nondyslexic ones:

1. I am very good at forming 3-D spatial images in my mind.

2. When I form 3-D spatial images in my mind, I can manipulate them at will and view them from all angles.

3. I have very good 3-D spatial reasoning ability.

4. I can create an image in my mind of the working parts of machines and how they operate.

5. I have always been good at building things. [For example, as a child: toys like LEGOs, robotics kits, trains, K'NEX, marble runs, models, 3-D arts and crafts projects; as an adult: things like landscaping, home repair or renovation, machine or engine work, furniture building, sculpture, etc.]

6. When assembling a kit (e.g., furniture, model), I usually don't have to read the written instructions. I can just tell how things must go together by looking at the pictures.

7. I generally prefer diagrams or pictures to written instructions or explanations.

8. I am very good at reading blueprints and at imagining in my mind the 3-D structures they represent.

9. When I think through a problem, my thinking is more nonverbal (visual images, other sensory images, movements, etc.) than verbal (words).

10. Before I can describe what I think about something, I often have to translate my thoughts into words (that is, up to that point, my reasoning process hasn't primarily used words).

11. I'm especially good at reading topographical maps and can easily imagine the landscapes they represent in 3-D.

12. After going someplace once, I usually don't need directions or a map again to find it or to find my way home.

13. I can picture in my mind the 3-D layout of places where I've been and walk through them in my mind.

For these M-Strengths questions, the sum of Strong Agreement + Agreement for the dyslexic respondents averaged around 72 percent, versus 24 percent for nondyslexics. Clearly this is a very substantial difference.

Look at questions 9 and 10, which deal with the tendency to think without words. During our statistical analysis, we found an extremely high correlation between the tendency to agree with the statements on spatial reasoning abilities and the tendency to endorse these statements on language. This strongly supports our observations in earlier chapters about the nature of this strength-challenge pair.

Next, notice how several of these questions deal with processing features that will be highly relevant for the classroom and workplace. See in particular question 7, on using diagrams or visual imagery for learning, and questions 9 and 10, on the difficulties of translating ideas into words. Both relate to the fact that images and not words are the primary mental language for many individuals with dyslexia. On the input side, many individuals with dyslexia will learn better when provided with diagrams, graphs, flowcharts, drawings, or visible enactments. On the output side, they can more easily communicate their thoughts when provided with opportunities to express their ideas and demonstrate their understanding through media other than words; extra time to answer questions in class or at meetings, or by informing them in advance what they will be called on to discuss, so they have some time to put their thoughts in words; or taking extra time to engage them in patient dialogue. In the other MIND strength sections, we will also consider how the answers to these surveys can provide helpful insights into best practices for school and the workplace.

New Research on M-Strengths in Dyslexic Individuals

Since the first edition of this book was published, evidence supporting the connection between M-strengths and dyslexia has continued to build. Dr. Matthew Schneps, whom we mentioned in an earlier chapter, was the lead investigator in two of these studies. Matt was for many years an astrophysicist, educator, and award-winning filmmaker at the Harvard-Smithsonian Center for Astrophysics. Throughout his career Matt was intrigued by the high prevalence of dyslexia he noticed among his colleagues in astrophysics. He was also intrigued by the sense that these dyslexic colleagues often seemed to have greater visual-spatial skills in certain areas than their nondyslexic peers. One such skill was the ability to detect things that seemed out of place or unexpected—especially things that lay outside the center of the visual field, toward the periphery. To determine whether his suspicions were correct, Matt and several colleagues performed two laboratory experiments.

In the first experiment, Matt and his colleagues compared ten struggling readers from Landmark College, a two-year preparatory college for individuals with learning and attention challenges, with nineteen good readers who were undergraduates at Harvard University. Each student was presented with a series of pictures that had been intentionally blurred using a technique called low-pass filtering. This process makes the whole picture appear roughly as blurry as the images we usually see at the peripheral portions of our visual fields, using the light receptor cells at the edges of our retina. (In contrast, the clear image that we typically see in well-taken photographs mimics the view we obtain using the foveal cells near the retinal center.) After the dyslexic and nondyslexic students were shown these low-pass filtered images, the dyslexic students were significantly better able to recall the layout of the objects in the images and to notice when changes were made. This result supported Matt's hypothesis that dyslexic individuals show advantages in certain visual and spatial skills—including the ability to detect and remember the positions of objects in space. Please note: These performance advantages result from a shift in visual attention from a more focused to a more diffuse pattern—a shift that is usually regarded as a sign of poor attention or distractibility. This is yet another example of why it is so important to consider differences as trade-offs rather than simply assume they are deficits.

In the second experiment, Matt and his colleagues measured the ability of dyslexic and nondyslexic scientists to detect radio spectral wave signatures of black holes. This task required the ability to see a subtle pattern formed by paired squiggles lying far to the periphery of the visual field on an instrument panel. The study found that the scientists with dyslexia were able to identify these target wave patterns significantly better than their typical reading colleagues. This finding once more shows how a practical benefit can result from a pattern of increased peripheral visual sensitivity that is most often regarded as a deficit.

In another recent study, Mirela Duranovic and colleagues compared dyslexic children to their nondyslexic peers using a paper-folding task that required complex mental simulation. The children were asked to predict how a piece of paper would look after it had been folded, then pierced by a hole punch, then unfolded again. The dyslexic children were significantly better able than the nondyslexic ones to make the correct prediction, demonstrating superior skills in mental spatial simulation.

Finally, since the first edition of this book was published, Kenneth Pugh and colleagues have replicated and extended the findings of Catya von Károlyi's previous two studies on the recognition of impossible figures, which we discussed earlier. As in those studies, Pugh and colleagues found that dyslexic individuals were quicker (but just as accurate) in identifying impossible figures than nondyslexic individuals. They also performed a second experiment in which dyslexic and nondyslexic participants were shown complex 3-D shapes, then asked to determine what the figure would look like when rotated. The dyslexic individuals were quicker than and just as accurate as the nondyslexic ones on this task as well.

Other New Research Relevant to M-Strengths

In chapter 8, we mentioned some of the exciting advances that have been made during the last decade in our understanding of the brain's spatial reasoning systems. Much of this exciting work has grown out of the research on the place and grid-cell system. To recap briefly, the brain is able to use this system to create 3-D models of real and imaginary spaces that allow us to plot our own locations in space, plan or imagine movements, or determine the spatial locations, positions, and structures of other objects and their component parts.

It should be clear from this description that M-strengths reflect the ability to use this system especially well. The reason individuals with dyslexia seem particularly likely to excel in this ability will become clear in later chapters, but very briefly it seems to result from the greater reliance dyslexic minds place on using lifelike episodic memories versus more abstract semantic memories to understand and remember things.

It has also more recently come to light that just as the grid-cell system in the hippocampus and surrounding brain structures can make three-dimensional models of space to support spatial reasoning, other types of cells in these same brain regions can collaborate to form multidimensional models of *nonspatial* concepts. To understand at a very general level how these systems might help us organize our nonspatial knowledge, let's look back at our discussion in chapter 3 where we formed a multidimensional model of a mountain expedition. As you'll recall, most of the dimensions that comprised that model weren't spatial at all (e.g., temperature, time of day, visibility). Yet they clearly had a "dimensional" character—often of a "more or less" nature—that was easy to recognize and understand. Those "more or less" dimensions are analogous to the spatial dimensions in our grid-cell network (i.e., height, length, breadth). The multidimensional model of dyslexia we have been forming is a nonspatial conceptual model as well.

With these examples in mind, we can begin to understand how our brains might form multidimensional models of concepts that resemble the spatial model formed by the place and grid-cell system. Using analogous systems, our brains can create mental models for countless kinds of complex concepts, then organize these concepts into networks, relationships, and hierarchies. As we have seen in our discussion of spatial reasoning abilities, the brain can use its spatial models to mentally simulate all sorts of past, future, and purely fanciful or imaginary positions, objects, or movements. And as we'll see in later chapters, it can also use many other kinds of conceptual models to help us organize our knowledge and facilitate our reasoning in countless other ways, many of which contribute to the other MIND strengths. Although this research has not been performed specifically on dyslexic individuals, we believe it has great relevance for understanding dyslexic strengths.

M-Strengths in Action: Beryl Benacerraf

When Beryl was only one year old, she moved with her family from America to France. Her father was a doctor and wanted to do research there. When they returned to America six years later, Beryl was just starting school.

At first, no one worried much when Beryl struggled with reading. They figured she was just confused because she had learned to speak two different languages. However, it soon became clear that something else was going on.

As Beryl recalls, "It was a nightmare for me at school, and I didn't do well. During reading periods I used to be sent down for special help, one on one, with an instructor." While these sessions helped to some extent, Beryl still could not read fast enough to keep up with her schoolwork.

Reading out loud was especially frightening for her. When her teacher asked the class to begin reading, Beryl would count how many students were in front of her, then try to figure out what paragraph she would be asked to read. She would spend the time until they reached her trying to puzzle out the words she would need to speak.

Beryl's struggles with reading intensified when she reached middle school and the books got longer. In seventh grade she was asked to read *Great Expectations* by Charles Dickens. Beryl knew she would never be able to finish such a long book, so she decided to watch a movie version of the story instead. She discovered that the film completely changed the story's ending only when she failed the class exam. (Ironically, the film was made by the great British director David Lean, who was also dyslexic.)

Despite her challenges, Beryl worked hard, and her high school grades were good enough to qualify her for many good universities. She knew, however, that she would not do well on the lengthy precollege test called the SAT (Scholastic Aptitude Test), so she applied to a university that would accept her without having to take the test.

In college, Beryl developed several important strategies that helped her succeed. She avoided classes that required lots of reading or memorizing dates, like history or literature. Instead, she took as many science and math classes as she could. When she had to write papers, Beryl dictated them and paid fellow students to type them up. These strategies enabled Beryl to do so well that she was able to get into medical school.

During her anatomy classes, Beryl realized she was very good in spatial thinking. She began to think she might want to become a surgeon. But before she applied for residency training, she took a one-month rotation in radiology. When that rotation was over, the supervising professor took her aside and asked if she had ever thought of becoming a radiologist. He told her he had never seen a student with her ability to spot patterns and abnormalities on radiology images. While other people focused almost entirely on small details, she took in whole images as single large patterns. He said this was a very great gift. Beryl considered his advice and eventually decided to become a radiologist.

That was an important era in the field of radiology, when many new imaging techniques were beginning to be used. One was called *ultrasound*. Ultrasound had a big advantage over conventional techniques like X-rays. It could make images using very high-frequency sound waves that would not harm sensitive body tissues like X-rays could. One type of especially sensitive tissue is the developing fetus. Early ultrasound machines could safely create images of developing babies in their mothers' wombs. However, the images made by the early machines were very hazy. As Beryl recalled, "The pictures looked like the surface of the Moon; but I was able to see what was going on." Beryl was so much better at reading these fuzzy images than most of the other doctors that she soon made many important discoveries. These discoveries have allowed better monitoring of fetal health during pregnancy, earlier recognition of genetic conditions such as Down syndrome, better diagnosis of many gynecological problems such as chronic pain due to endometriosis, and earlier detection of ovarian cancer. Eventually Beryl became not only a professor of obstetrics, gynecology, and radiology at the Harvard Medical School, but also the president of the American Institute of Ultrasound in Medicine.

Beryl firmly believes that her outstanding abilities to see and think in complex three-dimensional patterns are a result of her dyslexic mind. She told us, "I really think I owe my ability to see patterns in imaging to dyslexia. I'm very lucky to have this gift and to discover that I have it, because I've definitely struggled with the gifts that I don't have. But I've realized that there is a silver lining to being dyslexic. I believe that encouraging dyslexic people to find and develop the things they do better than most people is very, very important."

Part 4

I-Strengths

Interconnected Reasoning

13

The "I" Strengths in MIND

I-Strengths in Action: Jack Laws

Everything is about relationships. Things are as they are because of their relationships with everything else. You can't just look at anything in isolation."

As Jack Laws spoke to us by phone from his home in San Francisco, it became clear that his view of relationships and interconnection wasn't just a throwaway line but a true expression of his way of understanding and experiencing the world. It's a view he's shared with the growing number of readers of his amazing field guides on the wildlife of California, published under the name his parents gave him in tribute to another great California naturalist: John Muir Laws.

Anyone who's read Jack's field guides or attended his lectures knows that he's a teacher of exceptional skill. But as a child, Jack never expected he'd have anything to teach because he found it so hard to learn.

Jack was in the second or third grade when he first realized "there was something odd going on with my brain." Even though he was clearly bright and hardworking, he seemed unable to learn many things his classmates mastered easily, like reading, getting his letters to face the right way, and memorizing math facts.

Jack's parents took him for evaluation, and he was given a diagnosis of dyslexia. Targeted learning therapy greatly reduced—but didn't eliminate—his challenges with reading, spelling, writing, and math. "The therapists tried to tell my teachers how to help me learn, but the diagnosis of dyslexia wasn't a part of their training, and they weren't in a place where they could really hear it. This was new stuff, and I had teachers who thought that the best way to make me learn was to embarrass me more in front of the rest of the students—and that didn't really work for me. So I was bumped around to a number of different schools. I was smart enough to see that other kids could do this stuff, but try as I might, I couldn't, and over time I became increasingly convinced that I just wasn't one of the 'smart kids.'"

While Jack got little encouragement in the classroom, outside of school there were several "bright rays of light" that buoyed his spirit and encouraged the growth of his mind. "My dad was an amateur bird-watcher, and my mom was an amateur botanist, so on all of our family trips we would study nature together. I kept field notes of my observations, but it was difficult for me to write; so instead I made lots of diagrams and sketches of the birds that I would see and the flowers I would find and what the bugs were doing, and those sorts of things. Those notebooks became my way of training myself to look more closely at the world around me. If I wanted to sketch an object, I had to look at it again and again and again, and I discovered that even the most common things still held secrets and discoveries for me."

It was on those family outings that Jack first recognized that an intricate web of relationships connected everything in nature. "On these family trips, I was either down on my belly looking at flowers with Mom or finding birds with Dad or exploring around the mountain meadows, catching frogs. So I wasn't just out in flowers or out in birds: I was out in a place that had all of these different elements, and they were interacting with each other. So my sketchbooks were full of whatever I was looking at—not only bugs or only birds, but from a very early start, I was actually looking at ecosystems."

Jack's other "bright ray" was scouting. "I had a wonderful experience in Boy Scouting. I became an Eagle Scout and a leader in my troop, and I discovered that I could tie the fastest bowline knot in the San Francisco Bay Area Council. I was really good at leading a group of kids, solving problems, helping people get along with each other, demonstrating first-aid skills. And I really did well with map reading: I could look at a topographical map and the landform would three-dimensionally pop out at me, and I could route-find better than anybody else. That helped me have more self-confidence, which was a really important part of my childhood. But academically . . . school: not so much."

It wasn't until his sophomore year of high school that Jack finally began to find his way as a student. Appropriately enough his breakthrough came from connections he made with two of his teachers, one in biology and the other in history. Jack credits the profound impact they had on his life to the fact that they "stopped looking at my spelling and started looking at the content of what I was writing. There was a revolution in my brain because they helped me see that I had good ideas—I just couldn't spell them right. And when I finally realized that those were two different things, that was *huge* for me. It still took me forever to finish things, but for these two teachers, I would have done anything. Between the two of them, in one semester they turned me around. It was tremendously exciting. My life and the turns that it has taken are very much due to their influence."

While Jack considered careers in both history and biology, he realized, "If I went the history route, I'd have to read a lot more and write a lot more, but with biology, I could run around outside and listen to birds."

Jack's choice of biology—and his life's vision for his work—was finally (and fittingly) crystallized on a school trip when his class hiked the John Muir Trail. "On that trip I started fantasizing about a perfect field guide that would have full-color pictures of everything I was seeing."

For years the production of that field guide would remain Jack's cherished goal and the subject of countless daydreams. "I could visualize whole sections of the book and what the pages would look like and how I would organize it."

The next few years were a time of tremendous growth for Jack. "At the end of high school and going into college, I was getting the idea that there were different things I could do to compensate for the things I struggled with. Once you realize that you've got something really good inside you to share, then dyslexia becomes a much smaller obstacle, and you learn to compensate and deal with it." Recorded books and a word processor with spell-check were especially helpful in allowing Jack to succeed in school. "Not being afraid of technology and embracing what can help you is crucial. I still can't spell and I still don't know my multiplication tables, and I've never read a book from cover to cover without books on tape, but hey—it's okay! I'm doing just fine." "Just fine" for Jack includes a bachelor's degree in conservation and resource studies from the University of California at Berkeley, a master's in wildlife biology from the University of Montana, and a degree in scientific illustration from the University of California at Santa Cruz.

After completing this training, "I picked up my backpack and headed up to the Sierra Nevada and started painting wildflowers and animals. I didn't stop for the next six years."

We mentioned to Jack that the marvelous sense of connection with his subjects that he conveys in his field guides reminds us of the remarkable ability to "get inside" the minds of animals that we've seen many children with dyslexia display. "It's funny you should say that," he responded. "My connection with nature is not just an intellectual one. It's deeply spiritual and empathic. . . . When I give lectures about natural history, all these animals in my head are not just balls of factoids about this species or that species, but they're characters, they have personalities, I give them voices. Some people are very careful not to anthropomorphize the species that they see, and of course it's not scientifically correct to project a human perspective on all the things around you. But while I recognize that my human perspective is different than the world that's perceived by a wolverine, I still find myself constantly speaking for the wolverine or the pika and essentially trying to put myself in their shoes and give commentary about things from their perspective.

"I sometimes get this vision: I'll sit in a place, and I'll lean my back against a tree, and I'll look at something, and I'll think, *How does that relate to something else that's here?* Then I'll imagine a sort of colored line of energy going between those two things. Then I'll look at another thing and see how that relates to it, and I start trying to actually picture in my head the web in front of me of relationships between things: a three-dimensional web. And I start thinking to myself, *This is just some of the stuff that I've experienced or studied about,* and there are so many more that I don't know about or understand.

"John Muir said when you try to pick out any one thing by itself, you find it hitched to everything else in the universe. And the more we dig around as scientists and biologists, the more we see that that is absolutely true. You can't just look at anything in isolation. I don't know if seeing connections is easier for me, but sometimes I can actually visualize them. I just imagine the dense web of networks floating in front of me, and it's very clear to me that I'm a part of that web as well: those lines are connecting to me, and I'm connecting to those lines."

Interconnected Reasoning: A Web of Meaning

Jack Laws is an outstanding example of an individual with impressive I-strengths, or *i*nterconnected reasoning. Individuals with prominent I-strengths often have unique ways of looking at things. They tend to be highly creative, perceptive, interdisciplinary, and innovative in how they make connections and put seemingly disparate things together. No matter what they see or hear, it always reminds them of something else. One idea leads on to the next. We often hear from their parents, teachers, spouses, and colleagues things like "They just see connections that other people miss" or "She'll often say things that seem so off topic. Then five minutes later I'll finally get it and see that she jumped right to the key issue."

> I-strengths create exceptional abilities to spot connections between different objects, concepts, or points of view, and to recognize, understand, and reason about systems.

I-strengths include:

- The ability to see how phenomena (like objects, ideas, events, or experiences) are related to one another.

- The ability to see phenomena from multiple perspectives.

- The ability to unite all kinds of information about a particular object of thought into a single global or big-picture view.

You may wonder as you view this list how individuals with dyslexia could excel at making complex connections like these when they struggle to form connections that most people make easily—like the connections between sounds and symbols, or between basic math equations and their answers. The answer to this apparent paradox is easy: not all connections are alike. Some connections are simple, joining closely related objects or concepts. Others are more obscure, distant, or complex. I-strengths specialize in making these less obvious connections.

As we hinted at the end of the last chapter, I-strengths work by creating conceptual models that are analogous to the spatial models that power M-strengths. These conceptual models help us organize, understand, and manipulate information.

14

The Advantages of I-Strengths

The power of interconnected reasoning is its ability to link all an individual's knowledge, ideas, and mental approaches into integrated conceptual models with which we can understand, reason, and imagine. These models are incredibly powerful because they allow objects of thought to be approached from many different angles, levels, and perspectives, so they can be seen in new ways, related to other phenomena, and understood in a larger context. Three of the core skills, or I-strengths, that help to form these conceptual models are the ability to detect relationships between different ideas, concepts, or objects; the ability to shift perspectives or approaches; and the ability to reason using a global or top-down perspective.

Strength in Perceiving Relationships

The first I-strength is the ability to detect relationships between phenomena like objects, ideas, events, or experiences. These relationships are of two types: likeness (similarity) and togetherness (or association, such as correlation or cause and effect).

Relationships of *likeness* can link physical objects, ideas, concepts, sensations, emotions, or information of any kind, and the likenesses may range from highly literal to purely figurative. Some individuals with dyslexia excel at detecting only particular types of relationships (for example, those linking visual patterns or those linking verbal concepts), while others show more general strengths.

Several published research studies support the idea that individuals with dyslexia, as a group, show special talents for finding similarities and likenesses. In a paper published in 1999, English psychologists John Everatt, Beverley Steffert, and Ian Smythe compared the performances of dyslexic college students and their nondyslexic classmates on two tests of visual-spatial creativity. Both tests measured the ability to recognize similarities between different objects or shapes—that is, to see how one thing could represent or replace another. The first was an alternative uses task, in which the students were asked to name as many uses as they could think of either for soda cans or for bricks. The second was a picture production task, in which the students were asked to form as many different pictures as they could using five different geometric shapes.

The results were striking. On both tasks the students with dyslexia handily outperformed their nondyslexic peers, imagining 30 percent more possibilities on the alternative uses task and drawing over a third more pictures on the picture production task.

We noticed that many of the individuals with dyslexia we saw in our clinic showed remarkable strengths in detecting similarities among objects, structures, or physical patterns. Many were amazingly skilled at recognizing the works of particular designers, artists, or architects by their characteristic style, often before they were old enough to read. Many were also prolific inventors, builders, or sculptors. They often showed special skill in adapting whatever materials they found at hand to construct their projects, demonstrating an unusual ability to see analogies between these seemingly spare parts and other objects.

In one task that's a part of a common IQ test, examinees are given three sets of pictures, then asked to find one picture from each set that can be linked by a common concept. This task can unleash astonishing creativity among dyslexic examinees in our clinic. Though we used the same pictures for many years, we never stopped receiving entirely new answers to these questions from our dyslexic examinees.

Many individuals with dyslexia are also highly skilled in detecting similarities between different words and verbal concepts. These connections can include analogies, metaphors, paradoxes, alternate word or concept meanings (especially more distant or secondary meanings), sound-based similarities (like homophones, rhymes, alliterations, or "rhythmically" similar words), and similarities in attributes or categories (like physical appearance, size, weight, composition, and uses and functions).

During testing sessions in our clinic, we observed differences in how individuals with dyslexia understood and employed words and verbal concepts. As we discussed in chapter 4, when processing a word or concept, many individuals with dyslexia activate an unusually broad array of possible meanings. As a result, they are less likely to respond first with the primary or most common answers, and more likely to give unusual or creative answers or a range of possible meanings and relationships.

Dyslexic examinees were also more likely to detect more distant and unusual connections in tasks involving verbal similarities. For example, when we asked how *blue* and *gray* are similar, most people responded with the "obvious" answer that both are colors. In contrast, dyslexic examinees sometimes gave answers like "They're the colors of the uniforms in the Civil War" or "They're the colors of the ocean on sunny and stormy days." One very bright seven-year-old boy with dyslexia responded to our question "How are a cat and a mouse alike?" by announcing, "They're the two main characters on a highly popular children's television program called *Tom and Jerry*."

We also frequently observed this heightened ability to identify distant and unusual connections when we asked individuals with dyslexia to interpret ambiguous sentences—that is, sentences that could correctly be interpreted in more than one way. Ambiguities in word meanings are common in English because many words have multiple meanings or may be used as different parts of speech (noun, verb, adverb, etc.). In fact, the five hundred most commonly used words in the English language have an average of twenty-three different meanings *each* in the *Oxford English Dictionary*. Correctly identifying which meaning is intended in a particular sentence requires a processing system that's capable both of identifying many possible meanings and of choosing the appropriate meaning based on the overall context of the sentence. During our testing, we commonly asked examinees to find two or more meanings for sentences that are intentionally ambiguous, such as:

The chickens are too hot to eat.

I saw her duck.

Please wait for the hostess to be seated.

Students hate annoying professors.

They hit the man with the cane.

I said I would see you on Tuesday.

We often found that dyslexic students who had struggled with many other fine-detail language tasks were able to correctly interpret these sentences with no difficulty at all, while students who had excelled at fine-detail language tasks often struggled. Skill in recognizing alternate meanings is useful for interpreting all kinds of complex communications, like stories, jokes, conversations (especially informal ones), poems, and figurative language of all kinds (like analogies or metaphors). This skill is also highly useful for reading, especially for struggling readers.

Individuals with dyslexia who possess prominent I-strengths also frequently show an impressive ability to spot relationships of *togetherness* that link things, ideas, or experiences. This ability is sometimes referred to as *pattern detection*. Strength in detecting relationships of correlation or cause and effect is a useful skill in many fields, including science, business, economics, investment, design, psychology, leadership, and human relationships of all kinds. Jack Laws expressed his awareness of the pervasiveness of cause-and-effect relationships in nature in his description of the "three-dimensional web [of] relationships between things" that he often sees and in his recognition that "things are as they are because of their relationships with everything else."

Another dyslexic scientist who has demonstrated an acute perception of the interconnectedness of nature is Dr. James Lovelock. Lovelock is best known as the formulator of the Gaia hypothesis, which states that the climatic and chemical components of the earth's crust and atmosphere interact to form a complex system that maintains the earth in "a comfortable state for life." Lovelock was first led to posit such connections when he noticed subtle correlations in the variations of the chemical composition of earth's atmosphere and oceans. While other scientists before Lovelock had recognized that the earth's atmosphere was almost perfectly suited for biological life, none had realized that this special balance was maintained by the interactions of a tightly linked network of chemical processes. They had observed the same parts but had missed the interconnections that bind the whole system together.

Strength in Shifting Perspectives

The second I-strength is the ability to see connections between different perspectives, approaches, or points of view, and particularly the ability to shift between these different perspectives or points of view oneself. This I-strength also helps its possessors see that a particular problem, idea, or phenomenon can be studied using approaches and techniques that are borrowed from different disciplines or professions, and that subjects typically viewed as falling within a particular professional domain can be studied in meaningful and often innovative ways by individuals without the traditional professional credentials for that domain using new sets of perspectives and approaches.

Individuals with this I-strength prefer interdisciplinary rather than specialized approaches when taking on new problems or projects. They typically reject traditional ways of categorizing knowledge into "watertight" fields or disciplines and are dissatisfied with narrow and highly reductive approaches. Instead, they try to use as many different approaches as they can to solve problems and further their understanding, and they often borrow and adapt techniques from many different sources, applying them in new ways.

This way of viewing information often leads individuals with dyslexia to become "multiple specialists" who are knowledgeable in several fields rather than highly specialized in a single one. As a result of this interdisciplinary mindset, they often find new and creative ways to apply approaches from one field to others where they're not usually used. Sometimes their recognition that interesting questions require new approaches leads them to seek broader training. For example, James Lovelock already had a PhD in physiology when his growing interests in environmental science and climatology led him to pursue a second doctorate in biophysics. Ultimately, it was this blending of professional perspectives that suggested to him that the earth's biosphere might be understood and studied as if it were a physiological system.

Many individuals with especially severe dyslexic challenges have found themselves unable to complete their professional training through traditional academic routes. As a result, they have been forced to acquire their skills through work experience or self-education. John "Jack" Horner is a good example. When Jack was only eight years old, he began to find some strange-looking rocks in the hills near his home in Shelby, Montana. He soon realized they were dinosaur bones. Jack went to his local library to find information about dinosaurs, but he struggled to learn from the books. Unlike the other kids in his class, Jack still could not read even the simplest words. Although no one realized it then, Jack was dyslexic. Jack learned as well as he could from looking at the pictures in the library books, but he decided he could learn even better by finding more bones and studying them on his own.

Over the next few years Jack was able to find many bones. By the time he was twelve, his dinosaur bone collection was so good it was put on display in the local library. It is still there today. Jack kept finding more bones, and by the time he was seventeen, his dinosaur collection won the Montana high school science fair. As a prize, Jack was given a scholarship to the University of Montana.

Unfortunately, Jack still could not read or write very well. In fact, he had graduated from high school with the lowest grades possible. His principal had told him that he really should not graduate, but that they did not think they could teach him anymore, so he might as well leave.

During his first year in college Jack failed every one of his classes. However, in Montana at that time you could keep taking more classes even if you failed all your classes the previous term. So Jack kept trying, and he kept on failing. During the next five years he failed all his classes seven times.

Despite his poor grades Jack was a good listener, so he managed to learn a lot about dinosaurs. He also worked at several dinosaur digs, where he learned to recover dinosaur bones from the ground, and he worked at the university museum, where he learned how to clean and mount dinosaur skeletons for display. However, because he could never do well enough on tests to pass any classes, Jack never earned his college degree. Eventually Jack left school, and after working for a while driving a truck for the rock quarry in Shelby, he took an entry level job cleaning and mounting fossils at the Princeton University science museum in New Jersey.

On vacations, Jack went home to Montana and continued to dig for dinosaurs. Three years after leaving college he made an amazing find. Jack found a dinosaur nest. It was the first dinosaur nest ever found by anyone anywhere. It was a huge discovery, and it provided the first evidence that dinosaurs cared for their young.

Jack also found the first dinosaur eggs ever discovered in the Western Hemisphere. Although other scientists in different parts of the world had found dinosaur eggs before, no one had ever tried to see what was inside. They regarded the eggs as rare and valuable fossils. Who in their right mind would be willing to break one open to see what was inside? Well, Jack would. He was curious. He thought the eggs might contain fossilized embryos. He thought it was silly to have such a potentially valuable resource and not attempt to learn from it. After all, he reasoned, there was such a thing as glue. He could always stick the eggs back together after he'd broken them apart. So he took a hammer and broke some of the eggs; and when he did, he discovered the first dinosaur embryos anyone had ever seen.

Jack's discoveries revolutionized paleontology. He was given a MacArthur Foundation "genius award" to recognize his creativity. On his fortieth birthday, Jack even received an honorary doctorate from the University of Montana, the same school that had failed him seven times.

Now that he had a degree, Jack was finally able to become a professor and teach graduate students. Before long he had built the largest dinosaur research program in the world. He even wrote several books about dinosaurs, and in the early 1990s Steven Spielberg asked him to be the scientific adviser for the *Jurassic Park* films.

Jack is convinced that the same brain differences that make it hard for him to read and spell also help him see things other scientists miss. He told us he believes that dyslexic people are so good at "thinking outside the box" because they have never been inside the box. Jack thinks that rather than trying to fix dyslexic minds, we should try to understand how to use them better. That just might be his best idea of all.

Strength in Global Thinking

The third and final I-strength is the ability to combine different types of information into a unified big-picture or global view. This I-strength reflects the ability to perceive the whole that can be made by combining different parts and to identify its central essence. This critical skill is one of the chief components of the dyslexic advantage. We can better understand this big-picture reasoning strength by examining one of its key components: *gist*.

Gist is the essence or overall meaning of a thing, idea, concept, or experience. It's the "rough" bird's-eye view, rather than the fine-detail view: the forest rather than the trees. Gist detection helps us recognize context, so we stand a better chance of being able to fill in any information we find ambiguous or unclear and determine what's relevant.

All verbal messages have a gist, but this gist can't be determined by simply adding up all the primary meanings of each of the words in the message, then computing the global meaning like a sum. Instead, the gist or overall meaning of the message must be distilled by carefully considering all the possible meanings of all the words and phrases. Through this search for gist, clues about the source's meaning and intent can be identified, as can the source's mood and style. Ultimately, these gist-determining skills lie very close to the core of what we mean by *understanding*, and they're essential for determining the meanings of all but the simplest verbal messages and especially for complex messages like stories, plays, poems, jokes, or social interactions.

Gist also lends power to the other I-strengths. Gist detection allows us to determine the fitness of analogies and metaphors. It also helps us decide which perspectives, viewpoints, or approaches we should use to best understand some object of thought. In these ways, gist can be thought of as the ability to detect the relevant context or broader background of an object, idea, or message.

Researchers have shown that individuals with dyslexia tend to rely more heavily on gist detection than do nondyslexics for comprehending verbal information. T. R. Miles and colleagues at

Bangor University in Wales presented both dyslexic and nondyslexic college students with four sentences of increasing complexity and measured how long it took each student to master the verbatim repetition of the sentences. While none of the twenty-four nondyslexics needed over eight repetitions to master the fine details, the dyslexic students averaged considerably more trials and some required as many as twenty-five. Despite these difficulties in mastering the fine details, the dyslexic students performed just as well as the nondyslexics in summarizing what the sentence meant.

Many dyslexic readers are particularly dependent upon background clues and context to home in on gist. That's why many dyslexic readers show better comprehension for longer rather than shorter passages—*especially if the extra passages contain helpful contextual clues*. We have often listened as dyslexic students misread every second or third word of a complex passage and wondered how they would ever understand what they were reading. Yet when the passages provided background context, we often found that the comprehension of even severely slow and inaccurate readers was surprisingly strong, and sometimes outstanding. This kind of "upside surprise" was due largely to their ability to use contextual clues to grasp the gist of the passage, which allowed them to correctly guess at the identities of the individual words. In contrast, when passages contain few contextual clues, their comprehension usually worsens as the passages grow longer.

This beneficial effect of extra context is why many dyslexic readers enjoy books that are part of a series. Series books contain many of the same characters, settings, and activities and often use similar words. A similar improvement in comprehension can be seen when individuals with dyslexia are pre-equipped with a summary of the passage they'll be reading, shown a film version, or supplied with a list of key words.

This skill in gist detection is also very helpful in areas besides reading. Many individuals with dyslexia and prominent I-strengths develop a settled habit of searching carefully for gist and context in all areas of life. As a result, they often look for deeper and deeper layers of meaning and context beneath the obvious meanings. This pattern of continually peeling back the onion to find the deeper significance of an idea or thing or occupation is one we've seen repeatedly in the individuals with dyslexia we've studied. Often it plays a key role in their success, leading them to question things and ideas that have long been taken for granted, and allowing them to find secrets that have been hidden in plain sight.

I-strengths can create benefits for many kinds of tasks. Of course, like all of the MIND strengths, they can also be accompanied by challenges or trade-offs, as we'll discuss in the next chapter.

15

Trade-offs with I-Strengths

Each of the I-strengths we've discussed so far comes with its own set of flip-side challenges.

Trade-offs with Strength in Perceiving Relationships

While the ability to see broad fields of meaning is useful for detecting similarities, the tendency to spot more distant or secondary relationships rather than to fix immediately on primary connections can also worsen performance in certain settings. It's especially likely to cause problems in settings where speed, accuracy, reliability, and precision are more valued than creativity, novelty, or insight.

One setting of this type is the standardized test, including the IQ tests we described before. For example, in tasks like linking picture concepts or identifying verbal similarities (which we described in the last chapter), many individuals with dyslexia are so extravagantly good at coming up with insightful but nonprimary connections that we often find their test scores—which are based entirely on the number of right or primary answers they give—misleading.

Similarly, many dyslexic individuals that show I-strengths struggle with multiple-choice exams. Cynics might be forgiven for suspecting that multiple-choice exams—with their terse, dense, noncontextual sentences—were designed specifically to trip up individuals with dyslexia who excel in detecting secondary meanings or distant word relationships. These examinees will often pore over a multiple-choice exam like a lawyer vetting a contract, finding loopholes, ambiguities, and potential exceptions where none are intended. While their classmates evaluate questions using a standard of reasonable doubt, dyslexics search for proof beyond the shadow of a doubt. As a result, even a hint of uncertainty leads them to reject answers that most students would identify as correct. If their reading is also somewhat dodgy, the multiple-choice exam can become a nightmare. Even dyslexic students who read longer passages or whole books with excellent comprehension may struggle with multiple-choice exams.

This dyslexic talent for finding unusual connections can also lead to other difficulties in the classroom. Most of the tasks students are asked to perform in school—like reading fairly literal texts, responding to simple questions, or acting on straightforward instructions—are easier for minds that routinely fix on primary or common meanings. Students who call up more distant meanings can appear off target, especially if they don't get the simple answers that everyone else does or become confused by ambiguities no one else notices.

For some individuals with dyslexia, each word or concept may be surrounded by such a rich network of associations that these associations can become overwhelming and give rise to unintended substitutions. Sometimes these substitutions involve near-miss or similar-sounding words, like *adverse/averse, anecdote/antidote, persecute/prosecute, conscious/conscience, interred/interned, imminent/eminent, emulate/immolate*. While such errors are usually attributed to problems with sound processing (that is, difficulty with phonological awareness), a careful examination of dyslexic word substitutions suggests that factors other than impaired word sound processing are often also involved. Consider Mark, whom we saw in our clinic. Mark is a highly creative boy with a great fund of knowledge and a lively imagination. Yet he often struggles to say what he means. Sometimes his verbal substitutions involve words with similar sound structures, such as:

There were three people out in the missile. [middle]

I was looking at an add column. [ant colony]

Look at the winnows. [minnows]

Those people are cocoa. [cuckoo]

Being dizzy can really affect your carnation. [coordination]

That purple light caused an obstacle illusion. [optical]

At other times, though, Mark substitutes words that bear only a slight structural similarity to the intended word (e.g., sounds, length, "rhythm") yet share some relationship of meaning:

We made this for Dad's graduation. [celebration]

Max, quit ignoring me. [annoying]

Jim was there at the book club. [chess club]

At still other times, Mark substitutes words with almost no structural similarity, so that the relationships are purely conceptual:

Don't eat that—it will spoil your breakfast. [dinner]

That was a great Valentine's, wasn't it! [Christmas]

Mom, where's the bacon? [baloney]

Those curtains have polka dots. [stripes]

Conceptual substitutions like these are referred to as *paralexic* or *paraphasic* errors; and when made during reading they're sometimes called *deep substitutions.* While they are less common than sound-based errors, in our experience more individuals with dyslexia make them (at least occasionally) than is generally supposed.

In her autobiography, *Reversals*, dyslexic author Eileen Simpson vividly describes her frequent paralexic substitutions. One example she cites is her unintentional substitution of the word *leaf* for *feather*—a conceptual rather than sound-based substitution. Over time, Simpson learned to cover such slips (when they were pointed out to her) by pretending that they were intentional puns or jokes.

We believe this tendency to substitute related items is the flip side of dyslexic strengths in perceiving distant conceptual relationships. In support of this idea, we've found that the individuals who make these substitutions the most often excel on tests that require the ability to spot ambiguities or similarities.

Special strength in recognizing relationships of togetherness also comes at a cost. That's because skill at detecting correlations or causal relationships has been shown to be enhanced if your attention system is a little bit leaky or distractible. Many studies have found that individuals with dyslexia experience difficulty screening out irrelevant environmental stimuli, like noises, movements, visual patterns, or other sensations. This sensitivity to environmental distractions is one of the main reasons dyslexic students often need special accommodations for testing and other work that takes focused concentration. These distractions invade their conscious awareness and steal working memory resources.

At the other end of the distractibility scale, the ability to quickly and subconsciously distinguish between relevant and irrelevant stimuli so you can ignore what is irrelevant is called *latent inhibition*. Latent inhibition sounds like an unmixed blessing, and it's definitely useful in circumstances requiring tight attentional focus—like tests or silent work time at school. In fact, latent inhibition makes you the kind of student most teachers dream of having. However, before you conclude that students who test high in latent inhibition and low in distractibility are the lucky ones, you should know that there's an *inverse* correlation between latent inhibition (or freedom from distraction) and creativity. What this means is that the highest creative achievers tend to score lower on tests of latent inhibition. In fact, one study looking at Harvard students found that nearly 90 percent of those who showed unusually high creative achievement scored *below average* in latent inhibition—just like individuals with dyslexia. This is a critical fact to keep in mind when evaluating the balance between focus and distractibility in individuals with dyslexia.

Trade-offs with Strength in Shifting Perspectives

The second I-strength—the ability to shift between perspectives—is also remarkably useful, so long as you recognize when the shifts are taking place and they're under your control. However, we often find that individuals who can shift perspectives easily are subject, especially when younger, to switching perspectives without realizing it, and this can complicate certain tasks. For example, on a biology paper dealing with animal behavior, a student may begin by describing behaviors, then shift (appropriately) to a discussion of the neurological sources of that behavior, then veer off topic to consider other points of neuroscience that don't relate to the central topic. The student may also bring in elements of personal experience or opinion where they don't really belong and forget that they're writing a scientific treatise rather than an autobiography or opinion piece. Often such students' papers will have an air of free association that can be fascinating but takes them far from where they need to go. For students with strong perspective-shifting abilities, learning to control the team of horses that's tied to their mental chariot often takes great effort and prolonged and explicit training.

This ability to shift perspectives and to see things in interdisciplinary ways can create problems with organization as well. For example, high I-strength individuals with dyslexia are often horrific filers of papers. This is not, as is often supposed, simply because they have trouble alphabetizing but because they can think of too many places to file each paper and are more likely to lose papers that have been filed neatly away in distinct folders. As an alternative, they frequently prefer to keep papers in stacks, where they can more easily find them. There are some wonderful pictures that you can find online by searching "Einstein's office at Princeton" that beautifully illustrate this dyslexic "filing system." Fortunately, hyperlinked computer files and search capabilities have helped to reduce this problem.

Trade-offs with Strength in Global Thinking

The primary trade-off with strength in global or big-picture thinking is that it can create a greater dependence upon context and background information. Global thinkers have a top-down reasoning style that works best when a big-picture overview is already in place, so that new chunks of information can be added to conceptual frameworks that have already been built. That's why big-picture thinkers often learn best when they have at least a general understanding of the goals or ends at which they're aiming.

Big-picture, top-down learners are often a poor fit for the typical classroom, where bottom-up teaching approaches predominate. Schools often ask students to memorize new bits of information before explaining their meaning or significance. This approach doesn't really work for top-down learners because they can only remember things that make sense to them and new information that can be related to other things they already know. If they can't see the point of something they're asked to learn, it just won't stick. Without a big-picture framework to hang their knowledge on, the information is simply incomprehensible.

This dependence upon context is why "stripping down" instruction to the bare minimum to avoid overloading individuals with dyslexia often results in failure. Individuals with dyslexia who have a top-down, big-picture learning style typically learn better from approaches that convey information with greater conceptual depth, rather than from more superficial or survey-type approaches.

Individuals with this style also show several other characteristic patterns. For example, it's common to find dyslexic global thinkers at the upper levels of schooling still feeling lost far into the term, then suddenly finding that everything becomes clear when enough of the pieces are finally available to see the whole picture or build a functional conceptual model. Students with this pattern are often also more aware of (and bothered by) the gaps and deficiencies in the things they've been taught because they're more aware that parts of the picture are still missing. Students with this pattern typically do better the longer they stay in school. Upper-division college courses are generally easier for them than entry-level ones, and graduate school and postdoctoral work go even better than college.

For individuals with dyslexia who have this highly interconnected learning and conceptual style, a few simple steps can help them learn more effectively and enjoyably. For longer reading assignments, providing them with an overview beforehand (that includes both gist and context) of assigned passages can improve their reading speed, accuracy, and comprehension. If any new or special vocabulary will be included, giving them a list of key words also helps. Describing the practical relevance and potential applications of lessons will improve retention and motivation, as will tying in new information with things they've already learned. Finally, beginning each new course or unit by previewing the major points to be learned, and the approaches by which they'll be studied, can keep dyslexic students better oriented and more confident.

To show you how I-strengths and the challenges that go with them can appear at various stages of development, we'll turn in the next chapter to an individual with dyslexia who excels in interconnected reasoning. His name is Douglas Merrill.

16

Brains Without Borders

I-Strengths in Action: Douglas Merrill

When Douglas Merrill was young, he struggled to make basic academic connections. As he told us, "Reading was—and is—challenging, so getting through assignments meant using a bunch of tricks." With writing, "every other letter was backward." And with math: "Every summer my mother was reteaching me to add, subtract, multiply, and divide all the way up till I was in college. . . . Math never clicked for me. Even when I was in high school, I failed algebra."

Not surprisingly, Douglas's self-esteem suffered: "I always felt defective, which caused the sorts of things you'd expect in a kid, like super-defensiveness and hostility, because I felt like I was failing."

Douglas labored to get by on extra effort and sheer force of will, but the results were disappointing. It was only when he reached middle school that he realized that despite his difficulties with rote and fine-detail tasks, he also had special strengths.

One of his most important strengths was his ability to think and communicate using stories. "Pretty early on, I started writing stories to answer problems instead of doing what the assignment actually asked. So if you look at my junior high school papers, what I was doing for most of my classes was writing short stories. I was never going to be able to remember all the details that would be required to lay out a terse step-by-step outline. But I could remember the story arcs."

Douglas gradually realized how widely this technique could be used, but he was slowest to realize its relevance for math. "The breakthrough came in high school when I failed algebra. My dad's a PhD in physics, and my brother's a PhD in math, and one of my sisters is a practicing physicist working in nuclear power plants, so everyone but me is great in traditional math. When I failed, it forced me to look for something I did unusually well because I had to find some way to balance how awful I felt. I realized I could tell stories better than most people, and when I looked through what I had done wrong in algebra, I realized I was playing exactly to my weaknesses. I was not trying to make the math into a story. I was trying to memorize 'step A, step B, step C' by rote without creating any meaningful story about how they were connected, and that maximized my likelihood of failing."

Though school remained hard, Douglas began to make slow but steady progress. After graduating from high school he went on to the University of Tulsa, where he discovered a special fondness for studying the big ideas and forces that shape our world and for employing interdisciplinary approaches to those topics: "I have always been interested in the overlap between psychology, sociology, and history; the three work together to constrain what we can do, how we can do it, and how we view ourselves." Like many dyslexic individuals with powerful I-strengths, Douglas was unable to limit his focus to one subject, so he dual-majored in economics and sociology. In his spare time—which he claims to have had plenty of, as a self-described "can't-get-a-date geek"—Douglas picked up another skill set that eventually played a big role in his life: he became an expert in computer security and cracking computer programs.

After college, Douglas combined his interests in big-picture questions, learning and cognition, and computing by pursuing a PhD in cognitive science at Princeton, where he performed groundbreaking research on learning, decision-making, and artificial intelligence. When we asked him what motivated him to study cognitive psychology, Douglas responded, "Pure vindictiveness. I'd spent a lot of my life thinking through tips and tricks and different ways to solve problems, and I thought it would be interesting to think through how you could model that problem-solving—how to think about it and how to formally describe it."

As he studied these questions, Douglas realized that his insights into his own thinking could be useful for others, too. "When I investigated human problem-solving, one of the things I studied was people learning math and programming—partially out of a sense of irony—and I found that even normal people who are pretty good at the traditional rote skills do better if they form stories . . . even [for] things like writing programs. Now, no one I studied did this to the extent that I did, but I demonstrated a huge problem-solving improvement when you teach 'normies' to do what I did instinctively. That's something traditional schools could definitely teach to make their students more successful."

After earning his doctorate, Douglas went to work as an information scientist for the RAND Corporation, a prestigious think tank that researches public policy questions. At RAND, Douglas combined his expertise in decision-making and computer security to perform studies on computer security and information warfare—attempts by hostile entities to cripple an information system. Douglas found that his ability to think in stories and to see "the big picture" was incredibly useful in helping him imagine methods of attack that could exploit those weaknesses and to detect gaps in information security systems. As he told us, "RAND . . . is fundamentally a narrative place, so you're telling policy stories about what ought to happen, backed up by data, and that plays really well to my strengths. I can work with the people who do the data analysis itself, and I can say, 'Oh, here's where we're going.' I didn't understand that at the time, but that skill of being able to say, 'You know, I think we're going to head over there' was actually super useful—especially when I left to go into business."

Douglas's unique combination of skills eventually brought him to the attention of private-sector recruiters and to a series of high-level jobs at PriceWaterhouse, Charles Schwab, and eventually a small Bay Area start-up with the unlikely name of Google, where he served until 2008 as the chief information officer. Since leaving Google, Douglas has worked as president of digital music and COO of new music for EMI Recorded Music. He's now heading a finance company that he founded to help make bank loans more equitable and efficient.

When we were preparing to talk to Douglas, we noticed that he'd often used analogies to make his points in previous interviews. We asked him whether detecting similarities was a key element of his thinking. "Absolutely, I love analogies. They're my bread and butter—if you'll let me use an out-of-date analogy. Often the things I'm interested in doing arise from some analogy I come up with. I don't understand what I'm doing until I have a few analogies to describe it. The first element of my storytelling is asking, 'What's the story going to be like?' Then, for each major point, I brainstorm analogies."

When we asked him if he preferred to solve problems using interdisciplinary approaches and multiple perspectives, Douglas responded, "Yeah, totally. I just think there's lots of different views on problems, and that by seeing more than one you're better off."

Clearly, Douglas is equipped with remarkable I-strengths. As you'll see after you've read parts 5 and 6, he also has remarkable N-strengths and D-strengths. So what about M-strengths, or the kind of spatial reasoning ability that many people think of as the characteristic dyslexic strength? Douglas laughed when we asked him. "I'm abysmal at spatial reasoning. If I close my eyes, I can't tell you which way my office door is from where I'm sitting. But what I can do is tell you that it's nine turns from my office to my house, and they're on average three blocks apart. So I don't know where my house is from here, but I guarantee I can get you there."

Turns out that individuals with powerful I-strengths have many ways of making connections.

17

Key Points About I-Strengths

In the last few chapters, we've discussed the critical role that interconnected reasoning plays in the thinking of many individuals with dyslexia. Here are some key points to remember about I-strengths:

- The ability to spot important connections between various kinds of information is a critical—and possibly even the most critical—dyslexic advantage.

- I-strengths include the abilities to see relationships of likeness and "togetherness"; links between perspectives and fields of knowledge; and big-picture or global connections.

- I-strengths appear to be enhanced in individuals with dyslexia in part because their brain microcircuitry is biased toward the creation of highly interconnected long-distance circuits that favor top-down global processing and the recognition of unusual relationships.

- This structural and cognitive bias can lead to difficulties with fast, efficient, and accurate fine-detail processing.

- Dyslexic learners with prominent I-strengths can be greatly aided in their learning in various ways: for example, providing them with overviews of longer reading passages; helping them pre-learn key vocabulary; informing them about the practical importance of the material being taught; tying new information in with preexisting knowledge; and beginning units with a big-picture overview of goals and objectives.

I-Strengths in Action: Mimi Koehl

Let's close this section by looking at how I-strengths have helped another highly talented individual with dyslexia. Her name is Mimi Koehl.

The first thing Mimi can remember from childhood is a trip she took to the ocean. She was only three years old, but she still remembers vividly how fun it was to watch the waves. She loved the way they moved. She watched their patterns change in space and time. She followed them in and out, rising and falling, speeding up and slowing down.

Mimi was smart and creative, but she found school very hard. She couldn't memorize words or dates, and she struggled to learn to read. Even after she learned to sound out words, she read so slowly that she could never finish her assignments on time. However, by long hours of hard work she earned good grades in spite of her challenges.

Mimi's father was a physicist. He and Mimi spent happy hours at the kitchen table working out math problems together. Mimi thinks that he, too, must have been dyslexic because the tricks he showed her about how to keep equations from looking scrambled on a page are techniques that we now know help dyslexics. He also showed her how to build things in his workshop, like furniture for her dolls. Her mother was an artist, and sometimes Mimi would accompany her on hikes to lovely spots where they would paint landscapes.

When Mimi went to college, Mimi's parents pushed her to major in art, but because she went to a liberal arts college, she was required to also take classes in other subjects. In her biology class, Mimi learned that she was fascinated by the forms of plants and animals—not just from an artistic perspective, but from a functional one. That biology class changed Mimi's life. For the first time Mimi found a school subject she really loved. She came to realize that science was all about understanding and figuring things out—not just memorizing them. Even if she couldn't complete all the reading assignments or memorize all the word lists, she discovered that if she really understood the ideas behind all the examples that her teachers described in class, she could figure out how to answer their questions on the tests. After that class, Mimi decided that she wanted to become a biologist. Because she still loved the ocean, she decided to study marine biology.

Mimi became especially interested in the animals and seaweeds that grow on wave-swept rocky shores and coral reefs. These are some of the most important parts of the ocean. Like many dyslexic people, Mimi was especially good at looking at old problems in new ways and at combining different approaches in innovative ways to solve those problems. In her work on coral reefs Mimi combined research methods from separate fields like physics, engineering, and ecology—a classic I-strengths approach. Mimi was also able to simulate clearly in her mind how things move through space or in an ocean current and how they change over time in response to various forces or conditions. She discovered how corals alter the waves and currents flowing over them and how that water flow in turn affects other creatures living on the reef.

Before long, Mimi's new way of doing research made her one of the most admired ocean scientists in the world. The list of scientific prizes she has won for her work is remarkable, and like Jack Horner, the paleontologist we talked about earlier, she has received a MacArthur Foundation fellowship, nicknamed the "genius award" because it is given only to people who are especially creative in their work. In fact, it was only when Mimi met Jack at a MacArthur Foundation meeting and the two recognized that their minds worked in such similar ways that she finally learned to put a name to the challenges and the strengths that characterized her thinking.

Despite her great success, Mimi still struggles with certain things that other people find easy. When the University of California at Berkeley, where Mimi is a professor, switched from using keys to number pads to open the laboratory doors, Mimi found that she could never punch in the numbers correctly. No matter how great a scientist you are, you can't do your work if you can't get into your lab, so Mimi asked if she could have her key back. The university said no. Rules are rules. Fortunately, Mimi has a strong sense of justice and a talent for self-advocacy, so she told the school, "I'm dyslexic, and the law says I have a right to have a key if I need one!" The administrators were skeptical that one of their smartest scientists could be dyslexic and struggle with number pads, so they told Mimi if she wanted to be an exception and get her lock and key reinstalled, she had to spend a day getting tested to prove she was dyslexic. So Mimi did. And as a full professor at one of the world's top research institutions, Mimi finally got her official diagnosis of dyslexia. After she was tested, her psychologist told her, "Not only are you dyslexic, but you're one of the most dyslexic people I've ever seen!"—which is a phrase Mimi now recalls with pride. So Berkeley gave Mimi back her key, and she's still doing amazing work!

18

What's New with I-Strengths

In this chapter we'll focus on new information that helps to clarify the nature of dyslexic I-strengths. As in our update on M-strengths, we'll begin by looking at what dyslexic and nondyslexic individuals had to say when we asked them about their I-strengths.

What Dyslexic Adults Told Us About Their I-Strengths

The following are the questions we asked our pool of dyslexic and nondyslexic adults on our MIND strengths surveys relating to I-strengths. As with the M-strengths questions in chapter 12, the numerical data indicating strength of agreement can be found in the Appendix. Here are the questions:

1. I often see connections or relationships that other people miss (e.g., how things, people, events, etc., resemble each other, are alike, or are in other ways related or connected).

2. I often spot things or ideas that are missing or lacking (negative space thinking).

3. I am better at understanding the big picture than at thinking about details.

4. I often think of analogies and metaphors to better understand ideas or to innovate.

5. I tend to see relationships as complex webs rather than as links or chains.

6. I am good at detecting connections or relationships that make up a larger system.

7. I am good at detecting patterns in complex events or data sets.

8. I am good at approaching problems from different perspectives by switching my assumptions.

9. I am good at seeing things from another person's point of view.

10. I especially enjoy big-picture topics that try to explain the patterns behind what happens in the world, like science, philosophy, history, political science, theology, sociology, psychology, economics, etc.

11. I am especially good at spotting the gist or the heart of the matter in an argument, story, or set of events.

12. When I am taught a new task, I can't just learn it by rote, but have to make sense of it before I can learn it.

13. I learn best if I start with the big picture or general overview before trying to master the details.

14. When I study a new subject, I'm often confused about the importance of the details until it suddenly all clicks and I can see the big picture.

15. I often have difficulty learning something until I can understand its point, or why I am being asked to learn it.

16. I am good at coming up with new uses for things (that is, using them in ways for which they weren't originally intended).

17. I often combine approaches or techniques from different disciplines.

18. In my work I often use approaches or techniques that I came up with myself.

19. I often use techniques or approaches for applications that are different from those for which I was taught to use them.

On these I-strengths questions, our dyslexic respondents again showed very high levels of agreement, with the sum of agreement and strong agreement averaging around 86 percent for dyslexic respondents. This is an even higher proportion of dyslexic respondents than we saw with the M-strengths, though it is important to note that the nondyslexic group also agreed more strongly with I-strengths than M-strengths—in this case, 55 percent. Nevertheless, the dyslexic group still showed a significantly higher level of agreement than the nondyslexics.

The I-strengths questions in the list above can be clustered into three groups, each of which appears to indicate a related but slightly different ability. Questions 1 through 11 relate most closely to the core I-strengths concept of seeing connections, relationships, systems, and different perspectives. Questions 12 through

15 deal primarily with what we might call a top-down, big-picture, holistic, or connectionist learning approach. Questions 16 through 19 deal with interdisciplinary or innovative thinking.

Questions 12 through 15, which relate to an interconnected learning approach, clearly deal with issues that are highly applicable to both the classroom and the workplace. They emphasize the need to start with either a big-picture overview or a connection to something previously learned when introducing new topics. They also emphasize the need to provide background information. Many of the individuals we discuss in this book have commented on how important it was for their learning to have a firm base of context and how ineffective they found rote approaches. Appropriate approaches to helping students who show such preferences would include starting with stories, cases, or examples and then working down to particular facts or abstract principles and providing a sense of real-world significance of the topic under study. This top-down/interconnected learning approach should unquestionably be taken into account when teaching dyslexic individuals, as fully 19 percent of our dyslexic sample showed agreement with these items.

Questions 16 through 19 emphasize that many dyslexic individuals learn to do things in their own ways. Teachers, employers, and others should judge these different approaches by their results rather than their forms. If they make sense to the individual and they work, then flexibility is appropriate.

New Research on I-Strengths in Dyslexic Individuals

In the last decade, investigators in the field of creativity studies have produced some very important research bearing directly on dyslexic I-strengths. Creativity can be defined in various ways, but one aspect of creativity that lies close to the heart of every definition is *divergent thinking.*

Divergent thinking (also called *divergent creativity*) is the ability to come up with multiple possibilities or solutions to a given question or problem. It is the fuel for the creative fire in which new concepts, inventions, and strategies are forged. As a wealth of research has shown, divergent thinking is also a domain in which dyslexic individuals excel.

In earlier chapters we briefly mentioned the results of several laboratory studies that have looked at divergent thinking in adults. We won't repeat those references here, nor will we discuss each of the numerous additional papers that have been published on this topic since the first edition of this book was published. Instead, we would like to focus our attention on the results of two recent reports that have examined all these studies collectively using a process called a *meta-analysis.*

For those unfamiliar with the term, a meta-analysis is an aggregate study in which data from a large number of separate studies addressing the same question or questions are pooled, then statistically analyzed as a combined sample. The larger sample size created by this pooling process can sometimes help researchers detect differences between sample groups that were undetectable in the smaller-sized component studies.

In all, there were fourteen studies in one meta-analysis and twenty in the other. Both analyses returned similar results. No statistically significant differences in creativity were found between dyslexic and nondyslexic children or adolescents. However, dyslexic adults showed statistically significant increases in creativity relative to nondyslexic adults.

This adult creativity advantage was not limited to either the verbal or the nonverbal domain. Many of the component studies measured performance on common laboratory tests of creativity. On the verbal side, many used versions of the Alternative Uses Test. Dyslexic adults excelled on this divergent thinking task. Another component study in which dyslexic adults significantly outperformed nondyslexic ones examined the ability of participants to generate novel metaphors. On the nonverbal side, many studies used drawing tasks where individuals were asked to create figures, either starting from an abstract squiggle on the page or combining standard shapes like a circle or triangle to build their pictures. Again, the dyslexic adults excelled.

The authors of the meta-analyses considered several possible explanations for the apparently late-blossoming development of creative strengths in dyslexic adults. Most involved some form of compensation, as they theorized that these strengths were developed as compensations to make up for dyslexic deficits. Our interpretation is different. We view the strengths observable in adult dyslexic brains—including these strengths in divergent thinking—as *essential functions* of these brains. Rather than simply compensations or plan Bs, they are the abilities for which these brains are optimized through the structural and functional trade-offs we discussed in earlier chapters.

It should perhaps not be surprising that these creative strengths are more readily apparent in dyslexic adults than in dyslexic children given our earlier discussions about the changes that take place in dyslexic brains during childhood and adolescence. As we've discussed, network formation and skill acquisition are delayed throughout the early years, so many tasks continue to require conscious processing and working memory. However, working memory reaches peak levels only when individuals are in their mid- to late twenties, so minds that depend heavily on working memory for processing many tasks will likely take longer to reach their full power. As a result, dyslexic individuals may not be able to fully exploit their advantages until their working memory resources are fully mature. We could make similar arguments about dyslexic delays in development of processing speed, which also improve over time and continue even later into adulthood.

We will discuss the implications of this late-blooming nature of dyslexic development in more detail in later chapters. However, the lessons from these meta-analyses must certainly rank as one of the most important messages of this book for young dyslexic individuals: Don't lose hope when you begin to feel that you are the last tree in the garden to blossom. Your day is coming. Time really is on your side.

Other New Research Relevant to I-Strengths

Recent years have also seen many other new discoveries that are relevant to I-strengths. In this section we'll focus on studies relating to the ability to make creative connections like those we discussed in the previous section. To understand this research, it will be helpful to establish a little context.

One of the most often cited ideas in creativity research is the so-called *associative theory of creativity*. This theory was developed nearly sixty years ago by Sarnoff Mednick. Mednick proposed that high creativity reflects the ability to combine *remotely related concepts* in *highly original ways*. Since the concept of remoteness or distance is central to the associative theory, researchers must find ways to measure how remote or distant different words or ideas are from one another. In recent years, most neuroscientists have measured conceptual or semantic distance using methods first developed by researchers in linguistics and computer science. Words or concepts are considered semantically closer to one another if they often occur in the same passages, are used for similar purposes, or can be used interchangeably.

This method of measuring semantic distance can help us understand what the associative theory of creativity means when it refers to remotely related concepts. According to this theory, ideas are more creative when they *usefully* connect concepts that most people rarely if ever associate with one another. Creative people are those who excel at making remote connections.

This latter conclusion may have special relevance for individuals with dyslexia. Recall from our discussions in chapters 5 and 6 that dyslexic brains show greater reliance on right-hemisphere processing and more broadly spaced minicolumn networks. Both of these differences predispose people toward the ability to make precisely the kinds of remote and distant connections that we're now discussing. This suggests that dyslexic individuals may be optimized for making just these kinds of creative connections.

Of course, individual creativity depends not just on the ability to make *remote* connections, but also on the ability to make *many* such connections. As you probably remember from our discussion in the last section, the ability to make many creative connections is a core component of divergent thinking, and that is an area where adult dyslexic individuals really excel.

This raises a question: Why are dyslexic individuals so skilled in divergent thinking? One important clue can be found in a fascinating series of studies that have been conducted over the past decade, revealing a key link between divergent creativity and the use of episodic memory, or lifelike memories derived from personal experience. These studies have shown that when research subjects exercise their episodic memory systems right before testing, using a procedure called *episodic specificity induction,* they can significantly enhance their divergent thinking skills.

During episodic specificity induction, subjects are asked to think about some specific thing (like a person or place), then asked to recall as many vivid and lifelike details about some experience they have had with that thing for a minute or two. After this induction period, researchers have consistently found that subjects will for a brief time experience what researchers call an *episodic retrieval orientation.* This short-term orientation both makes them more likely to retrieve and use vivid episodic memories on whatever tasks they are given and significantly improves their performance on tests of divergent thinking.

We believe that this finding has great significance for dyslexic individuals. As we've already mentioned briefly, dyslexic individuals tend to store more of their memories in episodic forms such as experiences, cases, and examples, and to use their episodic memory systems for many types of reasoning tasks. This tendency will be our primary focus in the coming chapters on N-strengths.

I-Strengths in Action: Rebecca Kamen and Chuck Harrison

In closing, we're going to share the stories of two remarkable people whose life, work, and minds illustrate the value of I-strengths.

The first of these amazing individuals is Rebecca Kamen. Rebecca has a deep curiosity about the world around her and a gift for sparking curiosity in others. She is a natural teacher and a lifelong learner. However, she wasn't always a successful student.

"I had a lot of challenges growing up as a kid," Rebecca recalled for us. "I still can't read a book and remember it. Mathematics never made any sense to me because it was always just memorizing all this stuff and never connecting or attaching it to anything. When I finished high school in the late sixties I could not get into college—not a single one. I just couldn't do the SATs. It was awful. When I tell people I got 720 on the SAT, they say, 'Oh, was that on the math or on the English part?' And I say, 'No, that's combined.' So literally every college I applied to—including the community college in the town where I grew up—rejected me."

Still, Rebecca really wanted to go to college. Despite her challenges in the classroom, Rebecca dearly wanted to be a teacher. She felt that she had a lot of things to teach and that she could do it well. "It's not that I can't learn. I'm actually a very good learner in many ways. I like to learn with my hands, by taking things apart and studying them. I can also remember very complex things if I'm able to tie them in to something else and see how they relate to other things. That's the bottom line for me: if I can connect ideas through some story that makes sense to me."

Rebecca wasn't willing to give up on her dream, and she had determined supporters as well. "My parents were real advocates, and they knew how much I wanted to go to college. We also asked my high school principal for help, and he was willing to write a letter to Penn State." It worked, though some of the folks at Penn State were skeptical. "The counselor asked my parents why they were wasting their money sending me to college. In his estimation, I wasn't college material."

"Still," she told us, "they accepted me on probation for one year. They said, 'If you can pass the first year, then we'll admit you for good.' So I thought, *Wow, I've got to figure out a major where I don't have to take math, because that'll be the kiss of death.* And that's how I found art education. It was the only major at Penn State where I didn't have to take a math course. And as soon as I got into my first sculpture course . . . I found my way. When I started building things, I knew I was on the right path, because I always used to build things with my dad, though I never thought I could make a living out of it or turn it into a career. But from then on, school was amazing, and I did really well at Penn State. I got a fellowship to the University of Illinois for a master's degree in art education. Then after teaching for five years, I got a fellowship to Rhode Island School of Design, because I wanted to be a sculpture professor. And that's what I became. It's still amazing to me when I look back, because I know how challenging it all was."

Rebecca's teaching career was just one way she's excelled as an adult. She has also become a world-renowned artist with a rather unique focus: revealing the wonders of science and scientific creativity through art. Some of her most amazing artworks are on display at rebeccakamen.com.

One of Rebecca's recent exhibitions at the American University Museum bore the intriguing title *Reveal: The Art of Reimagining Scientific Discovery*. The magazine *Penn Today* described it in this way: "Kamen chronicles her own artistic process while providing a space for self-reflection that enables viewers to see the relationship between science, art, and their own creativity."

The profile in *Penn Today* also described an intriguing study by Penn researchers where Rebecca herself was the research subject. The study was led by Professor Dani Bassett, who has multiple appointments in engineering and neuroscience. One of Professor Bassett's many research interests is describing and measuring the curiosity and creativity displayed by individuals as their minds leap between ideas during various activities. One recent study looked at the mental leaps and connections people made as they browsed between topics on Wikipedia. Dani was so intrigued by the way Rebecca combined ideas in her artistic works that she asked Rebecca if her team could study how she made connections in her own mind. So Dani had the members of her team record Rebecca's descriptions of her thought processes as she creates her works of art. The team members used Rebecca's descriptions to build a data set of the subjects Rebecca discussed. They then constructed a 3-D visual network to display the connections Rebecca's mind made between concepts as she described her process of creating.

Dani and her colleagues found that Rebecca "tends to make large leaps between seemingly disparate ideas." This description recalls our earlier discussions about the relationship between creativity and conceptual distance. Another Penn researcher, David Lydon-Staley, described Rebecca's thinking style by saying, "This isn't just about information seeking. . . . This is jumping across networks and pulling pieces of information together to make something new, like a work of art. I think that's why Rebecca's art is so engaging—because it takes disparate concepts and creates something new and interesting out of them." Through the use of her amazing I-strengths, Rebecca uses her art to help others see those connections, too.

In addition to her work as an artist, Rebecca regularly lectures to students and scientists around the world on topics like enhancing creativity and seeing the world from new perspectives. Many scientists credit her artistic interpretations of scientific concepts with enriching their understanding of their own work. Helping others make new connections is one of Rebecca's driving passions.

The second person we'd like to introduce you to is Chuck Harrison. When Chuck was a boy, no one could understand why he had such a hard time learning to read. He was obviously very smart, and he seemed good at many things. Music. Sports. Building. Pretty much everything except reading.

Chuck still could not read by age nine. His parents took him for an eye exam and got him glasses, but these interventions did not help. Chuck's parents, his teachers, and his eye doctor did not know anything about dyslexia.

Over the next several years Chuck gradually learned to read well enough to get by, although his pace was very slow. Because Chuck was smart and a hard worker, by high school he was earning good grades. Yet when it was time for him to take his college entrance exams, Chuck did so poorly on the standardized test that the state university would not accept him.

At the time Chuck lived in San Francisco. The city had a two-year college Chuck could attend without taking an entrance exam, so he enrolled. Chuck still wasn't sure what he wanted to do for a living, so he took some economics courses. He almost failed out.

Chuck then went to the school guidance counselor for help. They gave him some tests. The counselor looked at the results and suggested that Chuck take some art classes, so he did.

To Chuck's surprise, he did very well in those classes. In fact, in his second year he earned straight As. Chuck's adviser thought he showed great promise as a painter or fine artist. However, Chuck preferred design or applied art. He especially liked the new field called industrial design. Industrial designers create plans for products that could be mass-produced for homes and businesses.

Industrial designers have to be skilled artists so they can make products that look good. They also have to know enough about engineering and mechanics that their products will work well. Chuck's special talent for combining ideas from different fields, like art and engineering, made industrial design a great fit for him.

Chuck decided to attend the School of the Art Institute of

Chicago (SAIC). At SAIC, Chuck's abilities quickly became apparent, and he was recognized as one of the prestigious school's top students.

After graduating from SAIC, Chuck began looking for a full-time job. He thought about working for Sears. It was headquartered in Chicago, and at that time it was one of the world's largest creators of industrially manufactured products. There was just one problem. It was 1956, and even though Chuck was already recognized as one of the best young designers in Chicago, Chuck was also Black, and Sears had never hired a Black person for such a high-ranking job. Chuck interviewed at Sears but was told by the head of the Sears design department that although he wanted to hire Chuck, the upper-level management at Sears would not let him hire a Black designer for a full-time job. Instead, he gave Chuck work as a freelance designer. This work helped Chuck get his professional start and allowed him to earn enough to live. Over the next five years, Chuck also worked with several designers whom he had met through SAIC. Gradually, the size and importance of the projects Chuck worked on grew. So did his reputation as a designer.

In 1961 Chuck received another call from the head of the design department at Sears. After years of badgering management, the department head had finally received permission to hire the amazing young designer. Chuck accepted the offer from Sears, and in doing so he became the first Black executive Sears ever hired.

Over the next thirty years, Chuck designed over 750 products for Sears. One of the things that made him so effective as a designer was his ability to view the products he designed not only from his own perspective as a designer but also from the perspective of each product's buyers, while also taking factors like material, labor, packing, and shipping costs into account.

Chuck eventually became the head of the Sears design department. During the years he ran the department, Sears was at the very peak of its success. Many of the products Chuck designed are still recognizable to anyone who lived in those days, including the View-Master, the first plastic garbage can, the portable hair dryer, and many more. It has been estimated that there were few households in America during the years Chuck worked at Sears that did not contain several products he designed.

Although Chuck passed away in 2018 and his name is not well known by the general public, he remains a legend in the design community. Chuck received many awards for his work, the best of which came in 2008, when he received the Cooper Hewitt, Smithsonian Design Museum's National Design Award for Lifetime Achievement, beating out all the other architects, graphic designers, and industrial designers alive at that time. This is the highest honor any designer can receive, and it is given to only one person each year. Chuck's amazing I-strengths, which allowed him to combine many skills and perspectives, made him one of the most admired industrial designers ever.

Part 5

N-Strengths

Narrative Reasoning

19

Narrative Reasoning: The "N" Strengths in MIND

N-Strengths in Action: Anne Rice

Anne was "a consistently poor reader" until well into adulthood. As is true of many struggling readers, her memories of school are highly negative: "School was torture. School was like being in jail. It was captivity and torment and failure." Though she dearly loved stories and spent hours flipping through picture books, her poor reading skills kept her from absorbing more than a bare sketch of the "action and incident" described on the page. Instead, it was through books read aloud at school and home, and the radio dramas and movies she enjoyed, that she developed a love for the rhythm and flow of language.

While reading was a struggle for her throughout elementary school, Anne's writing grew easier. From fifth grade on, she wrote adventure stories and plays for her classmates. They responded enthusiastically and overlooked her spelling errors.

Unfortunately, Anne found no way to turn her writing talent into classroom success.

It wasn't until her freshman year of high school that she finally read well enough to appreciate the actual words in the books she read. "The first novel that I recall truly enjoying and loving for its language as well as its incident was *Great Expectations* by Charles Dickens. . . . The other novel . . . was Charlotte Brontë's *Jane Eyre*. . . . I think it took me a year to consume these two books. It might have taken two years. [I]t was a slow go."

Despite these challenges, Anne's love of literature and writing continued to grow. When she went off to college, she decided to major in English. Unfortunately, she soon had to abandon this plan because she was still so "severely disabled as a reader" that she couldn't complete the assignments for her classes. Getting through even one of Shakespeare's plays in a week was virtually impossible for her, and the written work was equally difficult: "[I] barely got by . . . because I wasn't considered an effective writer. The one story I submitted to the college literary magazine was rejected. I was told it wasn't a story." Anne's spelling, too, remained a problem. As she told us, "I can't spell to this day. I don't see the letters of words, I see the shapes and hear them . . . I'm always looking up spelling and making mistakes."

Anne began looking for another subject where she might find more success. She was passionately interested in the great ideas and beliefs that shaped the modern world and wanted to form a "coherent theory of history." She considered majoring in philosophy, but here, too, she was hindered by her poor reading. Anne found that she "could only make it through the short stories of Jean-Paul Sartre, and some of the works of Albert Camus. Of the great German philosophers who loomed so large in discussion in those days [during the early 1960s], I could not read one page." Instead, Anne opted for a degree in political science, where she was able to grasp the key concepts almost entirely from lectures. She earned her degree in five years.

After graduation, Anne remained drawn to writing and literature. At age twenty-seven she returned to school to study for a master's degree in English, which she earned in four years. "Even then I read so slowly and poorly that I took my master's orals on three authors, Shakespeare, Virginia Woolf, and Ernest Hemingway, without having read all of their works. I couldn't possibly read all of their works."

Fortunately, Anne could still write, and shortly after earning her master's degree, she began work on a new novel. One of the primary themes of that novel was the experience of being shut out from life and the fulfillment of dreams—an experience Anne knew well from being "shut out of book learning." Three years later, that novel was published, and it became a phenomenal bestseller. Anne followed that first novel, which she titled *Interview with the Vampire,* with twenty-seven more, and together they've sold over 150 million copies, making Anne Rice one of the best-selling novelists of all time.

Dyslexic Novelists: More Common Than You Think

You might think it's extremely unusual for such a talented and successful writer to have trouble with reading and spelling. You would be wrong.

Many highly successful writers have faced dyslexic challenges with reading, writing, and spelling, yet have learned to compose clear and effective prose. Even if we limit our selection to contemporary writers whose dyslexic symptoms can be clearly confirmed, the list of successful dyslexic authors is impressive, and includes such notables as:

- Pulitzer Prize–winning novelist (*Independence Day*) Richard Ford

- Bestselling novelist (*The World According to Garp, A Prayer for Owen Meany*) and Academy Award–winning screenwriter (*The Cider House Rules*) John Irving

- Two-time Academy Award–winning screenwriter (*Kramer vs. Kramer, Places in the Heart*) Robert Benton

- Bestselling thriller writer Vince Flynn, all of whose novels were *New York Times* bestsellers and together have sold over 20 million copies

- Bestselling mystery writer, screenwriter (*Prime Suspect*), and Edgar Award winner Lynda La Plante

- Bestselling novelist Sherrilyn Kenyon (who also writes under the name Kinley MacGregor), whose novels have sold over 30 million copies

We're focusing on these talented writers because we believe they reveal something important about dyslexic processing in general—not just for dyslexic writers, but even for many individuals with dyslexia who never write at all. What these authors illustrate is the profoundly narrative character of reasoning and memory that is shown by many individuals with dyslexia. This narrative reasoning is the N-strength in MIND.

> N-strengths are the ability to construct a connected series of "mental scenes" from fragments of past personal experience (that is, from episodic or personal memory) that can be used to recall the past, explain the present, simulate potential future or imaginary scenarios, and test and grasp important concepts.

While many individuals with dyslexia might not instinctively regard their thinking as "narrative" in style, we'll show you the ways in which the memory and reasoning styles displayed by many individuals with dyslexia are in fact profoundly narrative. We'll also show you the many amazing ways that N-strengths can be employed.

20

The Advantages of N-Strengths

N-strengths draw their power from a kind of memory known as *episodic* or *personal* memory. We've mentioned episodic memory several times already in this book, but because it will be a main focus of this N-strengths section, let's quickly review what it is, and where it fits into our systems of memory and learning.

The human memory system can be divided into two main branches: *short-term memory* and *long-term memory* (see the figure on the facing page). Short-term memory—which includes the *working memory* we've already discussed—is responsible for "keeping in mind" the information you're using right now. Long-term memory, which will be our focus in this chapter, stores information you can retrieve and use later.

Long-term memory also has two branches: *procedural memory* and *declarative memory*. We've touched on both before. Procedural memory stores the procedures and rules that help you remember how to do things. Declarative memory stores facts

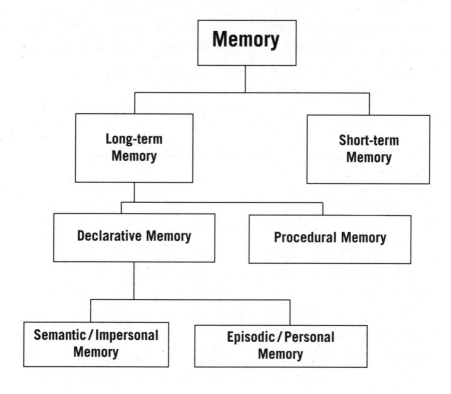

about the world. It is called declarative because it contains information you can consciously recall or declare.

Declarative memory can be further divided into *episodic* and *semantic memory*. *Episodic memory* (also called *personal memory*) contains factual memories that record either our real personal experiences or things we've imagined in the form of experiences. *Semantic memory* stores facts as abstract and impersonal data, stripped of context or experience.

Many facts about the world can be recalled either as episodic or semantic memories. For example, the fact that tears taste salty can be recalled as an episode or event you've experienced, or as a fact you simply know without remembering anything about the episode in which you learned it. We'll focus on episodic memory, because it underlies N-strengths, and because many dyslexic minds seem to have a special bias toward storing an unusually large proportion of their factual memories in rich episodic forms.

Understanding Episodic Memory

Episodic memory is the repertory theater of the mind. Episodic memories aren't stored as intact recordings in a single part of the brain—like old movies in a film vault. Instead, the visual, auditory, spatial, linguistic, tactile, and emotional components of episodic memories are disassembled, then stored in their respective processing areas throughout the brain—like stage props in a warehouse. Later, when an episodic memory is recalled, these "props" are retrieved from storage and reassembled into a form that closely resembles (or "restages") the original experience. To apply some of the terms we've used in previous chapters, each of these complete episodic memories comprises a part of our mental model of the world, connecting multiple bits of data and using grid-cell-like networks to link together elements like time, space, concepts, and all the other components of our lived mental experiences.

Like most dramas, episodic memories depict things that happen or are experienced, like events, episodes, or observations. They also contain traditional story elements like characters, plot, time, and setting. This gives them their narrative or storylike character.

This process of restaging mental scenes is an extremely powerful way of recalling facts about the world. We can get a glimpse of this power by inspecting the recollections of an individual with an extremely rich episodic memory, novelist Anne Rice. One of the most remarkable things about Anne's autobiography, *Called*

Out of Darkness, is the vividness and clarity of her memories from childhood, as shown in this description of a walk she often took with her mother:

> We left our house . . . and walked up the avenue, under the oaks, and almost always to the slow roar of the passing streetcars, and rumble of traffic, then crossed over into the Garden District. . . . This was an immediate plunge into a form of quiet. . . . I remember the pavements as clearly as I remember the cicadas singing in the trees . . . some were herringbone brick, very dark, uneven, and often trimmed in velvet green moss. . . . Even the rare stretches of raw cement were interesting because the cement had broken and buckled in so many places over the roots of the giant magnolias and the oaks.

This description is so clear—so rich in atmosphere and sensory detail—that the reader feels drawn right to Anne's side as she makes that journey. Yet when she wrote it, Anne was describing walks she'd taken nearly sixty years earlier.

As powerful as this "restaging" function is, recalling the past is only one of episodic memory's many functions. As the White Queen in Lewis Carroll's *Through the Looking-Glass* quite rightly observed, "It's a poor sort of memory that only works backwards." Episodic memory escapes this criticism because it does much more than help us recall the past. It also helps us understand the present, predict and envision the future, mentally simulate planned actions or inventions, imagine unwitnessed or fictitious events, solve problems, navigate, and create narratives that can persuade or enlighten others.

To help explain the many functions of episodic memory, we spoke with Dr. Demis Hassabis, a neuroscientist who played a key role in discovering how truly varied and powerful the functions of the episodic memory system really were. In 2007, Demis coauthored a groundbreaking paper with his colleague Dr. Eleanor Maguire. Their paper, which was voted one of the ten most important scientific papers of that year by the prestigious journal *Science,* introduced the term *scene construction* to describe the core process by which episodic memory works to perform its many functions.

When we spoke with Demis from his office in London, he explained episodic memory to us in the following way. "Episodic memory reconstructs things you've previously experienced from the remembered elements you've acquired through your experiences in life. For example, say you walk through a beautiful garden or park, and you see a beautiful rose, and you smell the rose: all those elements of experience become components in your memory. Later, when you want to recall what you've experienced, you reassemble those components in a way that looks familiar. You may get some of the details wrong because memory is often inaccurate, but to the extent that you're right, that's an accurate reconstruction of an episodic memory."

We then asked Demis to explain some of the additional functions of episodic memory. He responded: "Recently we've found that using scene construction to recall the past is just one small part of a much bigger system, which we call the *episodic simulation* system. Episodic simulation is very powerful, because it allows memory to be used creatively. With creativity you assemble the same kinds of memory elements that you use to recall the past, but rather than reconstructing something you've experienced before, you combine the elements in new ways to construct a whole that's entirely novel, because it contains unprecedented connections between the elements. In other words, creativity uses the same construction process that you use to reconstruct memories, but the construction is creative because it results in something you've never experienced before. The process is similar, but the outcome is entirely new."

This creative recombination system can be used as a kind of mental laboratory, to simulate what might happen given certain starting conditions or circumstances. According to Demis, "This episodic simulation function is very valuable in a lot of fields, including things like financial forecasting, or designing computer games and imagining how players will play them, or thinking about a film scene and how it might play out."

Recall that we mentioned earlier that many individuals with dyslexia appear to display a natural bias (or tendency) to store memories of factual information in *episodic* rather than *semantic* forms. We can now expand on that by saying that many dyslexic minds appear to show a natural bias toward using episodic systems like simulation and scene construction for more of their high-level cognitive processing than do nondyslexics, who often rely more heavily on more traditional logical or deductive processes. Dyslexic individuals with prominent N-strengths are especially likely to reason by mentally simulating potential events or actions, then observing how these simulations play out in their minds, rather than reasoning abstractly using definitions or formulas stripped of context. These simulations are based upon information they've gathered from real experiences rather than on abstract principles. This difference in cognitive bias has important implications that extend far beyond memory and reach to the very heart of the reasoning process, as we'll see in coming sections.

Scene-Based Versus Abstract, Noncontextual Knowledge

One way that the dyslexic bias in favor of episodic over semantic memory presents itself is as a tendency to store conceptual and verbal knowledge as *scene-based depictions or examples* rather than *abstract verbal definitions.* Often as a part of our testing sessions we asked individuals to define terms or concepts. While most individuals responded with abstract dictionary-type definitions, individuals with dyslexia often responded with examples, illustrations, anecdotes, or descriptions of uses or physical features. For example, when we asked individuals with dyslexia to define the word *bicycle*, they were more likely than nondyslexics to respond with an analogy (e.g., "It's like a motorcycle, but you make it go yourself") or a description (e.g., "It's a thing with a seat, two wheels, handlebars, and pedals that you make go by pushing the pedals with your feet"), as opposed to an abstract definition (e.g., "It's a human-powered two-wheeled transportation device"). The same was true when we asked individuals with dyslexia to define a concept that's inherently abstract, like *fairness.* Individuals with dyslexia were more likely to respond with an example ("It's like when you're playing a game and you wouldn't want to make someone else do something you wouldn't want to do yourself") than an abstract definition (e.g., "It means everyone should be treated the same" or "It means you get what you deserve"). This reliance on scene-based *depictions* of facts rather than abstract or noncontextual *definitions* reflects a greater reliance on episodic rather than semantic memory, and many of the adult dyslexic individuals with whom we've spoken confirm that this pattern is characteristic of their thought.

When we ask these older individuals with dyslexia to tell us more about their thinking style, they often also describe another feature that relates to episodic memory. When thinking of a fact or concept, they typically find that the concept is not represented in their mind by a single generalized depiction of that concept, but rather by a series of distinct examples through which they can mentally "scroll." The concept is understood as the complete collection of all these examples, and while it centers around the most common or representative examples, it also includes the outliers. Jack Laws gave us an especially good description of this conceptual style. When we spoke with him, he mentioned that he'd found he could much more easily distinguish between different animals of the same species than most people—for example, between different crows or robins. So we asked him what popped into his head when we mentioned the concept *robin*: Just one idealized robin, or a whole series of different robins? He answered without hesitation, "Different robins, definitely. My mind starts jumping to robins that I've experienced, rather than a single generalized robin or the Platonic ideal of the robin." When he drew a robin for his field guide, he drew the single robin that best represented the features of the whole group—but it was a *particular* robin, rather than an idealized generalization. In other words, he drew from an episodic rather than a semantic memory.

When we mentioned these dyslexic memory preferences to Demis Hassabis, he responded: "That's very interesting, because it relates to an important trade-off between episodic memory and semantic memory that people in the memory field have been thinking about. Both of these types of memory are critically dependent upon the hippocampus. But the interesting thing is—and this is really not written anywhere, but it's the sort of thing people at the cutting edge are thinking about at the moment—if you want to be very good at episodic memory, you want your hippocampus to engage in a process called *pattern separation*.

What pattern separation involves is this: Suppose you experience something new, and even though it's quite similar to something else you've experienced, you want to remember it as a distinct event—for example, what I did for lunch yesterday as opposed to three days ago, even though maybe I had lunch with the same people and in the same place, and lots of the elements were in common. One of the functions the hippocampus performs is to keep those similar memories separate. It actually makes them more divergent than you would expect—and that's exactly what you'd want if you want a good episodic memory.

"In contrast, if you want to be good at learning semantic facts that generalize and are true across multiple experiences—for example, the fact that Paris is the capital of France—then you don't really care about the specific episode you learned that in. What you care about is basically just the 'fact nugget.' The surrounding context in which you learned that fact simply isn't relevant to that piece of information. In that case, you want something else the hippocampus does, which is called *pattern completion*. Pattern completion is a process that unites divergent things. So let's say you heard a particular fact in several different lessons. What's actually important is *that fact*, not the different contexts in which you encountered that fact. So pattern completion solidifies your memory of the fact that was heard on each occasion, but eliminates any record of the differences in the way you experienced it.

"Now, if the hippocampus is responsible for doing both of those things, then perhaps what you're seeing with these dyslexics is that some of the same brain wiring differences that cause them to be dyslexic also predispose them to favor pattern separation over pattern completion. This [tendency to favor separated patterns] would make them very good at remembering things that have happened to them and at episodic memory. This greater diversity of separated patterns might also make them better at spotting unusual connections between facts that people who are not dyslexic wouldn't make."

This last observation is very important, because it relates to our discussion of I-strengths in the last section. There we suggested that creativity may be enhanced in individuals with dyslexia because they are predisposed to making more remote connections. If pattern separation also empowers the dyslexic mind by stocking it with more separated patterns from which remote connections can be made, then dyslexic minds might truly have a double helping of cognitive features that enhance their ability to make diverse and more creative connections.

Preserving the Uniqueness of Distinct Experiences

In the years since we first spoke with Demis, research has suggested a reason why dyslexic minds might be biased toward retaining more separated and lifelike episodic long-term memories, and away from forming more combined, generalized, and abstracted semantic memories. This reason appears to be based in the dyslexia-related challenges with procedural learning that we discussed in chapter 5. To understand this reason clearly, we must briefly explore how declarative memories are created and processed.

The earliest stage of declarative memory formation begins and coincides with the experience itself. Initially, sensory, emotional, and conceptual elements are linked together to form an *episodic* memory tracing of the event. For a short time, most of these episodic memories can be recalled to consciousness and replayed (that is, mentally reconstructed) in very clear and lifelike form. As time goes by, some of these detailed memories remain clearly accessible, others are forgotten entirely, and most are transformed into memory forms called *gist tracings*. Gist tracings contain a kernel that can convey a general description of the experience, but lack its lifelike character and details. They are episodic in the sense that they relate to specific events or experiences, but they lack the lifelike detail of fully episodic memories.

Consider this analogy. A filmmaker and a reporter witness a car crash. The filmmaker can capture the crash on video, then later replay the film for an audience, giving them sights and sounds in forms very like the original experience. In contrast, the reporter can only translate elements of the experience into the very different medium of language. No matter how vivid the description, the readers will not actually experience crash-like sights and sounds except in their imaginations. Clearly the video provides a more lifelike reconstruction of the initial event.

Over time, the gist tracings from many similar experiences are often combined to form *schemas*. Schemas are generalizations of the things we've learned from many similar experiences. Importantly for our story, schemas appear to be generated through a process that involves a form of procedural learning, which as we've seen is something many dyslexic individuals struggle with. Schemas form a major part of our *semantic memories,* which as we've discussed is the other main branch (besides episodic) of declarative memory.

Schemas are highly useful for quick and efficient reasoning and for storage of key facts and concepts. Given this description, you will probably not be surprised to find that they are largely stored in the brain's left hemisphere. Schemas lack the lifelike details that characterize episodic memories, and they also tell us more about the typical or usual than the exceptional or unusual.

This makes them useful for most routine situations, but less helpful in unfamiliar, confusing, or changing situations.

Now let's consider the importance of the observation we made just a moment ago, that this process of forming schemas appears to depend upon a type of procedural learning. If this is true, how might it affect people with dyslexia? We might expect to find, for example, that dyslexic individuals store and retrieve information in forms more closely resembling the memories of individual experiences—either as gist tracings or more lifelike episodic memories—than highly generalized schemas. We might find, particularly at younger ages, that dyslexic individuals appear overly focused on specific examples when asked to think about general concepts; yet they might also show great mental flexibility in their ability to think of exceptions to general rules. This flexibility might also cause difficulty in certain situations, like taking multiple-choice tests, where exceptions can be thought of for every generalization. Finally, we might expect to see enhanced creativity and problem-solving ability in new or unusual conditions, where memories of prior exceptional experiences would prove more useful than generalizations about common occurrences. These are in fact just the kinds of things that we do find when we examine the thinking processes of dyslexic individuals.

Together, the biases in dyslexic long-term memory that we've discussed in this chapter are ideal for producing minds with powerful narrative abilities. What could be more helpful for constructing narratives—whether complex novels or simple examples for a customer, student, client, or patient—than a mind stocked with particularities rather than generalizations; with an endless supply of different characters, experiences, examples, and scenarios; with a disposition to spot new connections, associations, patterns, and nuances; and with the ability to unite many different memory fragments into a single big-picture narrative?

Thinking in Stories: A Common Dyslexic Strength

This highly narrative thinking style often displays itself in a tendency to think and convey information in story form. We first noticed this tendency during our testing sessions when we asked individuals to describe a picture called the "Cookie Thief." This picture shows a woman standing at the front plane of the picture, drying a dish with a towel and looking out rather vacantly toward the viewer. Behind and to her left, water is overflowing from a sink where the tap has been left on, and it's beginning to collect in a puddle on the floor. To her right—and clearly unbeknownst to her—a young boy is standing on a stool, reaching high into a cabinet for a cookie jar that sits on the top shelf. Next to the stool, a girl is reaching up eagerly to receive a cookie. Neither seems aware that the stool is tipping, and the boy is about to fall.

Most viewers find the events in this picture rather trivial and implausible—especially the actions of the woman, who seems bizarrely detached from the chaos around her. As a result, they make little effort to reconcile the various events into a single coherent story. Instead, they simply describe the most obvious features of the picture. Over time, we found that a small number of viewers would propose additional details in an effort to reconcile the picture's seemingly irreconcilable elements. Most often the additional detail would involve a person or an object in front of the plane of the picture (like the father or a TV), which was distracting the woman's attention and causing her to ignore the sink and the children. Remarkably, nearly all the individuals who proposed these extra elements were dyslexic, and the solutions they proposed were clearly aimed at identifying the context or gist that would provide a coherent explanation for the action in the picture.

We also found that individuals with dyslexia were more likely to use conventional storytelling techniques to describe the picture. With younger dyslexics, these included formulaic openings like "One day while she was washing the dishes . . ." or "Once upon a time . . .". However, even older individuals with dyslexia were more likely to give the characters names, lines of dialogue, distinctive personality traits, senses of humor, motivations, and personal and family histories. We've found that many individuals with dyslexia use these kinds of narrative, personal, or episodic elements in all sorts of descriptive tasks, and that their descriptions often contain elements like analogies, metaphors, personalizations or anthropomorphizations, and vivid sensory imagery.

Strengths in narrative reasoning can also be seen in the professional lives of many individuals with dyslexia. The following are several examples of dyslexic individuals in fields other than creative writing who have flourished using narrative skills.

Duane Smith is a professor of speech and the director of the public speaking team at Los Angeles Valley College—a school he once failed out of as a student due to his dyslexia-related challenges. As he told us, "My whole life has been about stories, and telling stories, and I stress the importance in my public speaking class about telling stories. Half of what we do in forensics competitions is to perform stories, but for me, literally everything is about stories." Prior to becoming a professor, Duane had a successful career in sales, where he also found his narrative skills to be invaluable.

When we described episodic memory to Duane he laughed in recognition: "If I hear a song or smell something or see an article of clothing or a car from a particular year, I can immediately imagine a scene on a particular day, or event. It drives my wife crazy, because we'll be listening to the radio, and I'll talk about how it takes me back to 1985 when I was standing with a group of buddies at In-N-Out Burger on a Saturday night listening to that song, and what we were talking about, and she'll say, 'Can't you ever just listen to the song?'" By contrast, Duane told us that he remembers almost nothing in abstract, noncontextual form: "The only things I remember are experiences and examples and illustrations."

Law professor David Schoenbrod recalls that his parents were told by his English teacher when he was a junior in high school, "David is literate in no language." That's a problem he's long since overcome, as readers of his many highly regarded books on environmental law will confirm. To what does David attribute his success as a litigator? As he told us, "It seems to me that my strength as a lawyer was being able to tell a story. I had a colleague early in my career who told me that the way you win a case is by telling a story in a way that makes the judge want to decide your way. And I've always felt that I was good at that. . . . I like storytelling, and it came readily to me."

Entrepreneur and cognitive psychologist Douglas Merrill, whom we met in chapter 16, attributes his survival in school and his mastery of math techniques primarily to his use of stories. "I always think in stories. . . . I spent most of my time [as an adolescent] reading or telling stories, or playing fantasy games around stories like Dungeons & Dragons.

"I ended up at Charles Schwab, and Charles—who goes by Chuck—is dyslexic, and he sits in meetings with his eyes closed listening to people talk, and he never reads the handouts in advance, and it's pretty clear that all he's doing is listening and thinking. And he tells these great stories about what customers are going to want. I found that incredibly freeing, because for the first time I thought, *This stuff I do well at is valuable, as opposed to the stuff I do badly at, which seems to be what everyone else thinks matters.*"

After leaving Google, Douglas worked briefly as president of new music at EMI. "I thought one of the problems that industry had was that it didn't know anything about itself. So I spent a lot of time trying to figure out what's actually happening in the music industry. The most direct way to do that is with math and I suck at math, so instead I would read economic articles and surveys. I would make notes on yellow stickies, then stick them on the wall. Then once a week I'd skim those stickies and move them around, and what I ended up with was a story."

In short, N-strengths can be useful in any job or task where past personal experiences can be used to solve problems, explain, persuade, negotiate, counsel, or in some way form or shape the perspectives of oneself or others.

21

Trade-offs with N-Strengths

In addition to these many abilities, N-strengths can also bring trade-offs. The most important trade-off—and the most common—is reflected in the following comment: "Sammy never remembers anything from school. He forgets what's been taught, and whether there's an assignment or a test. And when we ask him to do something at home, it's always in one ear and out the other. His memory is horrible! Only . . . the strange thing is, Sammy is also our family historian. He can remember what we've done on every vacation, and who gave what present at his brother's birthday party five years ago, and what kind of pet every kid in his class has. So why can he remember all those kinds of things, but he can't remember his times tables or the names of the state capitals?"

This classic but seemingly paradoxical description of the "family historian with the poor memory" is one we hear from countless families.

All individuals, whether dyslexic or not, show a distinctive blend of strengths and weaknesses in episodic, semantic, and procedural memory. This blend greatly affects their learning and memory styles. Like Sammy, many individuals with dyslexia have relatively strong episodic memories, weaker semantic memories, and weaker rote and procedural memories (as discussed in chapter 5). Dyslexic individuals with this memory style are typically good at remembering things they've done or experienced, and often at remembering stories they've heard or information that's been embedded in a narrative context. However, they're much weaker at remembering "bare facts"—or facts that are abstract, impersonal, and devoid of context.

Anne Rice was a perfect example of an individual with this memory pattern. As we mentioned earlier, Anne had a phenomenally good memory for episodic and personal details of things she'd experienced; yet Anne also had a poor memory for abstract, impersonal facts. As she told us, "I don't think abstractly at all. Everything is image and narrative with me. I can't remember numbers at all, and make huge errors, sometimes doubling prices or amounts as my memory of them gets hazy."

It's critical to identify students with dyslexia who show a primarily episodic memory and processing style, because their N-strengths can provide the key to unlocking their learning potential. This is true both for the ways they take in and the ways they express information.

When taking in new information, students with dyslexia who show a strong episodic memory bias and a narrative processing style will typically learn much better if general or abstract definitions are supported by scene-based examples or depictions. When information is embedded in a context that the student finds meaningful and familiar and which incorporates experiences, cases, examples, stories, or personal experiences (including humor, participation, novelty, and "strangeness"), they will learn it more quickly and retain it more durably.

These points are reinforced by the following experience, shared by the nondyslexic mother of a dyslexic child. She told us that when she was a student, she'd always excelled at remembering facts, definitions, and formulas, while her dyslexic husband had always struggled in those areas. She'd been an honors student, while her husband had barely made it through school, so she naturally assumed she'd be a better tutor for their dyslexic son. She was surprised to discover, then, that her husband was a *much* more effective teacher, especially for concept-rich subjects like history, social studies, and science. Eventually she realized why her "poor student" husband was such an effective teacher. He taught almost entirely using examples, cases, stories, and analogies, while she tried to "trim the fat" from her lessons and present only the bare minimum of "simple facts" so their son would have less to memorize. But it wasn't actually the *quantity* of information their son was struggling with. It was the *form*. He could hold on to facts embedded in a meaningful context or story, but quickly forgot facts lacking background, connections, or significance.

Dyslexic students with narrative reasoning styles also face important challenges when trying to express their ideas. Since their conceptual knowledge is often stored as cases, images, or narratives rather than as abstract principles or definitions, when asked to answer questions on exams, assignments, or even orally in class by stating the relevant abstract or general principle, they may respond instead with stories or examples. As a result, their answers may appear loose and unstructured. They may seem to "talk around" their answer and appear to have difficulty "getting to the point." Douglas Merrill shared one example of how this happened to him.

"When I took my qualifying exams in graduate school, I was asked a question about the development of cognition. I was supposed to start with Piaget, then go to Erikson, then go to modern cognitive problem-solving; and although I understood all the important concepts, I hadn't really been able to memorize all the little details, so instead I wrote a story about the different developmental paths of two people. I covered all the right concepts, but they failed me on that question because I didn't give them all the specific details that they wanted."

This dyslexic tendency to think in examples or stories rather than in abstract definitions can also result in the loss of points on standardized tests, including (and perhaps especially) the vocabulary portion of the IQ test.

Schools and exams often treat abstract facts and principles as if they were the only forms of knowledge that really count, and assume that if students can't memorize and regurgitate facts in their "purest" and most noncontextual form they don't really *know* them. While abstract definitions are important and useful, we must not undervalue knowledge embedded in experiences, stories, cases, or examples. Such *case-based knowledge* is both highly valuable in its own right and more easily mastered than abstract information by many students with dyslexia.

It's also important to recognize that individuals with a largely narrative reasoning style will often show a very different trajectory of cognitive development from individuals with a more abstract or semantic reasoning style. This is particularly true of the growth of their conceptual knowledge. At younger ages, individuals who store concepts as examples have fewer experiences to reason with, so they may seem "stuck on" overly specific cases when asked to think about a broad concept. Early on, such children often have more difficulty generalizing their knowledge than their peers. Fortunately, as their experience increases, so will the fluidity of their thinking. In fact, once they have accumulated a broader set of experiences, they will often be *less* concrete than others, because their concepts include a wide range of cases rather than a single generalized principle. This also makes them less likely to mistake an abstraction or generalization for a full description of reality.

Finally, because narratives speak so powerfully to individuals with dyslexia who possess prominent N-strengths, it's crucial that the narratives we tell them about dyslexia are both accurate and appropriately hopeful. One of our chief goals in writing this book is to correct the common and deeply misleading narrative that dyslexic differences are primarily, or even entirely, dysfunctions. The story we should read in the lives of the individuals with dyslexia isn't a tragedy: it's an exciting narrative filled with hope, opportunity, and promise for the future.

22

A Story for Every Occasion

N-Strengths in Action: Blake Charlton

Let's look in detail at the many different uses to which one highly talented individual with dyslexia has put his N-strengths.

Blake Charlton was diagnosed with dyslexia midway through first grade. Despite being passionately fond of hearing and telling stories, Blake made little progress with reading and writing, and he struggled with basic math.

Blake spent two years in a special ed class, but then things slowly began to click. He enjoyed feeling like the "smart kid" in class and was pleased when he could finally start putting stories down on paper.

By fourth grade Blake had made enough progress to be mainstreamed back into the regular ed class. However, his sense of accomplishment quickly vanished as he went from being the smart kid to the class failure. Only his skills in sports and drama—which earned the admiration of his classmates—allowed him to keep a positive self-image. They also helped him respond to his classroom setbacks with a determination to improve.

While Blake was in middle school, his determination to improve helped him boost both his reading speed—largely by consuming the fantasy novels that gripped his imagination—and his pace of work. However, Blake continued to make "silly" mistakes in writing and math, and they chipped away at his grades.

Blake finally made a breakthrough when he was allowed to use a calculator and a spell-checker for his work. His grades shot up, as did his self-esteem. "Suddenly I was a geek again!" Blake enjoyed being recognized for his intelligence, and he began to apply himself with even greater determination. With accommodations in place for his College Board exams, Blake did well—so well, in fact, that he was admitted to Yale University.

Although Blake has fond memories of Yale, he also remembers college as a time of terror. He was so concerned that his remaining dyslexic difficulties would defeat him that he compensated by spending "every waking moment of the day" on his studies. Fortunately, Blake received invaluable assistance from Yale's Resource Office on Disabilities. They helped him obtain classroom accommodations such as a keyboard for in-class work, extra time on tests, and out-of-class assistance with proofing papers, planning, and scheduling.

Because he'd long dreamed of becoming a doctor, Blake began taking courses in science. He discovered that he had a special knack for chemistry. Blake particularly excelled in organic chemistry, with its heavy emphasis on three-dimensional spatial reasoning, and he actually earned the top grade in this very difficult and competitive class. Blake also did well in inorganic chemistry, because even though his rote memory was weak, he was still able to remember an astonishing number of facts about the chemical elements in the periodic table by creating fanciful stories about them. He gave each of the elements personalities, past histories, motivations, and goals, and these narratives helped him remember their "behaviors" and their positions in the table's rows and columns. Blake used similar narrative-based memory strategies in his other courses, too. As a result of these successes, Blake told us that for the first time in his life he felt truly "intellectually talented."

Blake might have been happy as a chemistry major had his love of stories not been so strong. Instead, he majored in English, and with persistent effort he saw his writing skills begin to blossom. In fact, he won two writing awards while at Yale.

Following his graduation in 2002, Blake took a job at a school as a combined English teacher, learning disabilities counselor, and football coach, then returned home to care for his father, who was battling cancer. In the few spare moments he somehow managed to find, Blake wrote stories about the imaginary worlds he'd dreamed of all his life, while continuing to dream about becoming a doctor.

In 2007, Blake finally entered Stanford Medical School. That same year he also signed a three-book deal with a publisher specializing in fantasy fiction. Blake's first novel, *Spellwright*, was published in 2010. Fittingly, *Spellwright* is the story of a dyslexic magician in training who must solve the riddle of his own cacography—his inability to handle text-based spells without corrupting them—in order to prevent the triumph of evil over good. It's an absolutely thrilling read, and the elaborate system of magic Blake created is astonishing in its inventiveness.

When we first spoke with Blake, he was taking time off between his second and third years of medical school to teach creative writing to first-year medical students, as a way of encouraging their use of narrative reasoning in clinical medicine, and to publish an analysis of literary narratives related to medicine (such as Tolstoy's *The Death of Ivan Ilyich*). During our conversation, Blake told us how useful he'd found narrative-based memory strategies for dealing with the overwhelming amounts of memorization he'd faced in his first two years of medical school. He even shared several of the stories he'd developed to help remember the branches of arteries and nerves. Narrative reasoning is clearly still a dominant theme in his life.

In the years since the first edition of this book was published, Blake has gone on to publish the second and third volumes of his Spellwright trilogy, completed a residency in internal medicine and a fellowship in cardiology, taught and published important writings on literature in medicine, and received the Pinnacle Award from the International Dyslexia Association in recognition of his role in raising awareness of the incredible potential of dyslexic people. Clearly in Blake's life, magic doesn't only happen between the covers of a book.

23

Key Points About N-Strengths

Narrative reasoning plays a key role in the thinking of many individuals with dyslexia. Key points to remember about N-strengths include:

- Many individuals with dyslexia show a profound difference between their powerful episodic (or personal) memories for events and experiences and their much weaker semantic (abstract or impersonal facts) and procedural memories.

- Episodic memory has a highly narrative or scene-based format, in which concepts are conceived or recalled as experiences, rather than as abstract definitions.

- The episodic construction system can use fragments of stored experience not only to reconstruct and remember the past, but also to imagine the future, solve problems, test the fitness of proposed inventions or plans, or create imaginary scenarios.

- Individuals who rely on episodic or narrative concepts will typically reason, remember, and learn better using examples and illustrations rather than abstract concepts.

- Many individuals with dyslexia enjoy (and are skilled in) creative writing, even though they may have difficulty with formal academic writing or reading; so teachers should look carefully for signs of narrative ability in students with dyslexia and should help talented individuals further their abilities through the use of appropriate tutoring and accommodations.

- Narrative approaches can be useful for all sorts of occupational and educational tasks, not just creative writing.

N-Strengths in Action: Kevin Horsley

Blake Charlton provided a good example of ways to use story-based memory strengths to remember all sorts of information, like the chemical elements in the periodic table or the branches of nerves and blood vessels. But in case you think Blake's story is unique—or the kind of thing that applies only to doctors from the Ivy League—we want to introduce you to Kevin Horsley.

By the time Kevin was eight years old, it was clear he could not learn the way most of his classmates did. Kevin struggled with reading, made very little progress with writing, had a hard time focusing on his work, and could never seem to remember anything.

Kevin was sent to the school psychologist. After testing Kevin, the psychologist gently explained to Kevin's parents what he thought he had learned. The psychologist said that Kevin had a kind of brain damage. It was called dyslexia, and it meant that

Kevin would always have difficulty reading, writing, remembering, and concentrating as well as other people. The psychologist recommended that Kevin go to a special class for students who could not learn well.

Nothing the school tried seemed to work well for Kevin. He could barely read his textbooks, so the only way he could pass his classes was by getting his parents or friends to read his lessons to him, then trying to memorize what they said. Kevin remembered just enough to pass his classes, but he never understood much of what was read to him—and remembering things that barely made sense was hard. By the time he graduated from high school, he still had never read a book from cover to cover. In fact, he barely read better than he had in first grade.

For the first few years after high school Kevin had little hope for his future and no real idea of what to do with his life. He began to believe that unless he improved his reading and his thinking generally, he would never get anywhere in life.

Kevin decided to do something he had never done before. He visited a bookshop. Kevin bought three books, but the one that most gripped him was called *Use Your Memory*. That book began a process of change that soon revolutionized both Kevin's memory and his entire life.

Kevin learned as he read that memory is not a gift that only some people have. It is a skill and a habit that can be improved with training and practice. Kevin was thrilled and inspired by this message, and he read with a dedication and a focus he had never before given any book.

As Kevin struggled through the book, he began applying and practicing its methods. He soon saw improvements not only in his memory but also in his mental focus and reading. This progress made Kevin more determined than ever to read and apply everything he could find on learning and memory.

Within a few months of finishing his first book, Kevin had read dozens. This reading practice improved Kevin's reading speed dramatically. Applying his new memory strategies to his reading improved his comprehension and retention, too.

Through his reading, Kevin discovered that the reason he had struggled with remembering things was not because his memory was simply bad. Instead, he had been trying to use a technique called rote memory that did not work well for him. Rote memory is not actually a very strong memory strategy for anyone, though many people can use it well enough to get by in school. But it is especially weak for most people who are dyslexic, and if they try to rely on it, it will almost always let them down.

Kevin learned instead that the most powerful memory strategies are based on skills like creativity, imagination, storytelling, making mental connections, remembering past experiences, and using three-dimensional spatial imagery. Like many dyslexic people, Kevin had excellent skills in these areas.

For the next four years Kevin worked hard to build his memory—harder than he had ever worked on anything before. Even he was astonished by how much progress he made. Kevin was so inspired that he did something that might sound crazy. He left his home in South Africa to compete against the world's best memory experts in the World Memory Championships in England.

Only a few years before, Kevin could hardly remember his phone number. Now he shocked everyone by taking fifth place in the World Memory Championships. Kevin's dyslexic mind proved no handicap at all. In fact, his exceptional creativity, connection-making skills, memory for past experiences, and spatial imagination all helped him remember with astonishing skill. And lest anyone think that this connection between dyslexic thinking and outstanding memory ability is hard to swallow, it is worth noting that the World Memory Champion that year was named Dominic O'Brien. As a child, Dominic had also been diagnosed with dyslexia, and he would eventually win the competition a record eight times.

Following the competition Kevin continued to build his memory skills. Four years later he set a World Memory Record in a test that has been called "The Everest of Memory Tests." For this test, Kevin had to memorize the first 10,000 digits of the number pi (π = 3.1415 . . .). After the 10,000 digits were broken down into groups of 5, the examiners would randomly call out the numbers in one of these groups of 5 digits; then Kevin had to respond by naming the next 5 digits on either side of that group. Kevin had to do this fifty times correctly, which he did; but Kevin also did it fourteen minutes faster than anyone had ever done it before.

Since that time, Kevin has traveled around the world, teaching people how to build their memory skills and to learn more effectively. This once poor reader has now written several books, including the outstanding *Unlimited Memory,* which describes the strategies Kevin used to transform his own learning and memory. And Kevin has done all this not despite, but with the help of, his amazing dyslexic mind.

24

What's New with N-Strengths

In this chapter we'll focus on new information that helps to clarify the nature of dyslexic N-strengths. As in our updates on M-strengths and I-strengths, we'll begin by looking at what dyslexic and nondyslexic individuals had to say when we asked them about their N-strengths.

What Dyslexic Adults Told Us About Their N-Strengths

The following are the questions we asked our pool of dyslexic and nondyslexic adults on our MIND strengths survey relating to N-strengths. Note how many of the questions deal specifically with memory strengths and challenges, and how many of the strengths match the attributes we've described as belonging to episodic memory. As with the M-strengths and I-strengths questions presented earlier, the numerical data indicating strength of agreement can be found in the Appendix. Here are the questions:

1. When I try to recall a fact or a concept, the first thing that comes to my mind is usually a case, example, personal experience, or image, rather than an abstract verbal definition.

2. When I'm trying to learn a new concept, I prefer my instructor to start with a story, case, or example, rather than to start with a precise definition.

3. I typically reason, remember, and learn better using stories, cases, or examples, rather than abstract concepts or definitions.

4. When I recall my past, I don't just remember naked facts about who, what, when, where, why, and so forth, but I reconstruct my old experiences in my mind in such vivid detail that it's almost like I'm experiencing them again.

5. I often need to use special tricks to memorize rote information (naked facts or lists), like making up a story, song, or rhyme about them.

6. When I think of historical facts or events, I see scenes or depictions in my mind, rather than just recalling verbal descriptions.

7. If someone asked me to think about an abstract term like *justice* I'd usually think first about stories or images or examples of just behavior, rather than abstract definitions of justice, like "giving everyone their due" or "treating everyone the same."

8. If someone asked me to define *tree* or *car*, the first thing my mind would do is imagine a whole series of trees or cars, rather than jump directly to an abstract, word-based definition.

9. I usually remember things I'm told in one-on-one conversations much better than things I hear in lectures or read in books.

10. I'm much better at remembering things I've done than things I've simply been told about.

11. I think I notice and remember more things that I encounter by chance in my environment than most other people I know.

12. I often seem to remember things I encounter by chance experience better than the things I tried hard to memorize in school.

13. I learn much better by practical experience than by formal instruction (e.g., books, lectures, paperwork).

Our dyslexic respondents agreed with these N-strength statements almost as strongly as they did with the I-strengths statements, with about 85 percent agreeing or strongly agreeing, versus 51 percent for the nondyslexics.

Questions 1 through 8 deal primarily with the form or type of material remembered. As we've discussed, dyslexic individuals tend to favor using memories with an episodic form, such as cases, examples, and stories, both for remembering factual information and for reasoning about concepts. The implications of these results both for learning and expressing information should be clear. When teaching students with dyslexia, using cases, examples, and stories rather than simply providing fact nuggets stripped of context or real-world details is essential to getting new information to stick. On the output side, showing greater flexibility in allowing students to express understanding through examples or depictions rather than simply parroting precise definitions is also highly appropriate. Of course, in those situations where the precise definitions must be taught (for purposes of passing a certifying exam, for example), remember that the precise definition will stick better after the student has first learned to understand the gist or essence through a few cases and examples, because this will provide something to hang the abstract definition on.

Questions 9 through 13 deal primarily with what we might call the learning environment. These answers show how much better individuals with dyslexia retain information that is encountered in the form of a personal experience, rather than information packaged in the form of a formal lesson. Questions 11 and 12 deal also with a phenomenon called incidental memory, which we'll turn to in the next section.

New Research Demonstrating N-Strengths

Look again at questions 11 and 12. Notice how more than twice as many dyslexic as nondyslexic adults agreed with each, and three to four times as many strongly agreed. Both of these statements relate to a type of memory known as *incidental memory*. Incidental memories are long-term memories that are acquired in the course of activities that are either nondirected or are aimed at some goal other than studying or learning about the object that was remembered. Incidental memories are like random bits of experience that stick to your mind as you wander through your day, and incidental learning (the process of acquiring incidental memories) is another special strength of dyslexic individuals.

Incidental memories are often stored in the form of lifelike episodic memories. In chapter 8, we mentioned an experiment that demonstrated that dyslexic college students showed superior incidental memory to their nondyslexic peers, though we did not use the term incidental memory at that time. It was the experiment in which Elizabeth Attree and colleagues asked dyslexic and nondyslexic college students to "search" through a 3-D computer simulation of a house with the goal of finding a toy car. After the students found the car, the computer monitor was shut off, and then the students were asked to do two things that were not a part of the original assignment. First, they were asked to reconstruct the floor plan of the house using architectural blocks. Second, they were asked to mark the location in that floor plan of a second toy that was not the one they were told to search for. Successful responses to both tasks required the use of incidental memory, and the dyslexic students significantly outperformed the nondyslexic ones on each.

In 2013, Michael Ullman of Georgetown University, along with colleagues in Sweden, performed another study to compare the incidental learning abilities of dyslexic and nondyslexic students (average age eleven). Michael is one of the world's leading authorities on procedural learning and memory, and he had previously observed that individuals who struggle with procedural memory—including those with dyslexia—seem to excel in declarative (or conscious) forms of memory. Because incidental memory is a type of declarative memory, Michael suspected that dyslexic individuals might show advantages in this area.

In their experiment, Michael and his colleagues showed children a series of drawings. Each drawing depicted either a real or an imaginary (made-up) object. Prior to showing the students the drawings, they told the students that their sole task was to say whether each object pictured was real or imaginary.

The dyslexic and nondyslexic students turned out to be equally good at this task. But that's when the *real* experiment actually began. Ten minutes after the students looked at the pictures, the researchers showed them a new set of pictures. About half of this new set were pictures recycled from the first part of the experiment, and about half were new. Instead of asking, "Is this object real or imaginary?" this time they asked, "Did we show you this picture before?" What they were testing was incidental memory—that is, whether the students had, without having been asked, incorporated these images into their long-term memories well enough to recognize which pictures they had previously seen. The results confirmed Michael's hypothesis. The dyslexic students were significantly better at recognizing which pictures they had and had not seen. When they repeated the same task twenty-four hours later, the dyslexic students were still significantly better than their nondyslexic peers.

These are just a few of the experiments that reveal that dyslexic people do not in any general sense have poor memories. Rather, they show "sawtooth" memory profiles, characterized by strengths in some forms of memory and weaknesses in others. Dyslexic individuals often show poorer performance in areas like auditory-verbal working memory and procedural memory. Yet they also show considerable strengths in various forms of consciously accessible (or declarative) long-term memory, like episodic and incidental memory. That is why it is so important to understand the memory strengths and weaknesses of each individual with dyslexia, so that strategies can be adopted to optimize performance on tasks that require the retention of information. As we saw earlier with Blake Charlton and Kevin

Horsley, and as we'll discuss in the next section, those strategies should focus on using the typically powerful episodic or personal memory system that dyslexic minds possess.

Other New Research Relevant to N-Strengths

The Thoroughly Narrative Nature of the Dyslexic MIND

In the years since the first edition of this book was published, research has revealed many new uses to which the episodic memory system can be applied. This work has dramatically expanded our understanding of episodic memory's importance not only for N-strengths but for all the MIND strengths. It is now clear that all four MIND strengths are all much more closely and inextricably linked than we previously imagined. We can now see that in the first edition we were like the four blind men in the parable who believed they had discovered a leaf, a snake, a wall, and a tree, when in fact they had found the ear, trunk, side, and leg of a single elephant.

The four MIND strengths are like different aspects of a single "cognitive elephant." At the brain's network level, they are all supported in large degree by the same three systems. We have discussed two of these systems in detail: the episodic memory system itself and the hippocampal grid-cell system. The third key system is called the *default mode network*.

The default mode network plays a critical role in many important cognitive functions, but it was discovered entirely by accident when functional brain imaging was first being developed. Researchers noticed that before their formal experiments began—in the moments when their subjects were simply left on their own to lie quietly in the scanners and think their own thoughts—certain brain regions remained consistently active and in communication with one another. These regions included centers in the frontal attention areas, centers in the right hemisphere linked to episodic memory and other higher cognitive functions, and centers in the back part of the brain linked to mental imagery. The researchers concluded that these parts of the brain formed a network that became active by default when the brain was not given any other task to do. Hence, default mode network.

Soon, however, it became apparent that this network was not just activated when the brain was merely relaxing or hanging out. Instead, it became engaged during any brain activity that involved an internal, rather than outward-directed, focus. These activities included recalling past experiences, recombining those past experiences in imaginative or creative ways, or engaging in a wide range of complex reasoning and problem-solving tasks.

As an example of a task in which the default mode network plays a vital role, think back to the study that we cited earlier comparing the brain regions used by expert versus novice physicians when they were trying to solve complex clinical problems. The brain regions that were preferentially activated by experts were in fact precisely those belonging to the default mode network. One of the key ways the default mode network acts is by using the bits of information and mental models provided by the episodic and hippocampal grid-cell networks (along with the other centers mentioned above) to engage in all sorts of complex higher cognitive processes. Many of these processes also contribute to D-strengths, so we'll discuss them in more detail in the coming chapters.

What we'd like you to focus on now is specifically how episodic memory plays an important role in default mode network activities, just as it does in the other two key systems underlying the MIND strengths. The key point is this: that the MIND strengths engage episodic memory in so many ways that it is now impossible to miss how thoroughly episodic and narrative—in the sense that they restage reality—the MIND strengths are.

Think back to our discussions of I-strengths. As you'll recall, we mentioned that dyslexic adults typically display superior performance in divergent thinking to nondyslexics; however, anyone's creativity could be increased for a short time by engaging in so-called episodic induction exercises—that is, by intentionally exercising one's episodic memory retrieval systems for several minutes. We speculated that the dyslexic superiority in divergent creativity might arise because dyslexic minds, due to their bias toward engaging episodic memory systems, were essentially *always* engaged in episodic induction exercises and therefore always boosting their divergent creativity.

Now, imagine if this sort of induction phenomenon could enhance not only divergent creativity but also any other mental activity supported by episodic memory—including all of the MIND strengths. Couldn't that also lead to a dyslexic advantage in each of these areas? We believe it could, and we are not alone. As creativity researcher Kevin Madore has written regarding the above research on divergent creativity, "an [episodic induction procedure] that biases a specific retrieval orientation *should likewise facilitate performance on subsequent tasks that nominally involve . . . episodic processing for completion* [italics ours]." In other words, a bias that creates more extensive use of episodic memory systems in general should enhance their performance for any of their applications. As we have seen, these applications include scene construction, model building, mental simulation, the activities of the default mode network, and all of the MIND strengths. Whether one is reasoning about three-dimensional space or about complex concepts like justice or freedom, or creating a story, or painting a picture, or engaging in many other kinds of complex tasks that we'll discuss in the section on D-strengths, all of the MIND strengths engage the core networks that employ episodic memories. Although their end products differ because each also engages different kinds of episodic memories and additional processing centers, the core systems powering each are highly similar, much as a single type of battery can power many different machines. Because dyslexic minds appear to be biased toward more frequent use of episodic memory systems, they should enjoy an advantage in the performance of all these applications.

This focus on the highly narrative and episodic nature of the dyslexic mind should also cast a clarifying light on the challenges that dyslexic individuals experience in schools and workplaces. Many of these challenges arise because episodic thinkers are being asked to function in highly semantic environments—by engaging rote memory; learning rules, abstract or generalized facts, and verbal definitions; and expressing their understanding in equally abstract verbal forms. The way forward for dyslexic minds is to learn how to nurture and engage the powerful episodic systems, which they naturally favor and which have so many valuable applications, rather than to try to function within the semantic straitjacket.

As a closing example, let's turn back to businessman Gerry Rittenberg's frequent question, "What is dyslexia?" It should be obvious by now that what Gerry was looking for was an episodic-based multidimensional mental model that he could use to *see* dyslexia, rather than an abstract verbal definition that simply told him what it was. While verbal definitions can provide a certain kind of superficial factual knowledge, they often do so without conferring any deep understanding. In contrast, a richly episodic mental model that accurately portrays all the relevant physical or conceptual dimensions provides both a knowledge of facts and a flexible understanding that will be of immense value, particularly in changing or unique circumstances. This kind of deep and flexible understanding is the ultimate end product of all the MIND strengths.

N-Strengths in Action: Philip Schultz and Larry Banks

Let's close this chapter by looking at the stories of two remarkable individuals. Both make their livings using narrative skills, and their lives display the many broad applications of N-strengths and of the episodic simulation system we've just discussed.

When Philip Schultz was eleven, he had still not learned to read or write, so his school had him see a tutor. After several lessons the tutor asked Phil what he wanted to do with his life. Phil said that he wanted to become a writer. The tutor began to laugh. He laughed so hard his belly shook and tears came to his eyes. When he finally stopped guffawing, the tutor asked Phil, "Why do you want to do something for the rest of your life that you've found so hard to do on its most basic level?" Phil did not know how to answer him, but he knew what he wanted to do.

When that tutor looked at Phil, he saw only a boy who could not read, could not write, could not tell time from a clock, could not learn left from right, and could not understand and follow directions. But the tutor did not understand dyslexia. He did not understand that for many people with dyslexia, things most people find easy can be hard, and things most people find hard can sometimes be easy.

Phil's teachers did not understand dyslexia, either. They thought Phil was simply unable to learn. When they called on Phil in class, he stuttered badly and would sometimes become so anxious that he would just get up and leave the classroom. Phil was often placed in a corner and just left to himself, as if he weren't really part of the class. One time a teacher put a book in his hands and said, "Here, look at the pictures. Just sit there and pretend like you're reading it."

The other students often picked on Phil because of his problems learning. Since he could not respond in words, he often responded by fighting back. Twice he was expelled from school, and he was made to repeat third grade.

Phil's parents and grandmother supported him, but some of his cousins teased him. Phil was Jewish. As in many Jewish families, the children in Phil's family went to Hebrew classes to study the scriptures they would need to recite to become adult members of their community. Phil was completely unable to learn the texts. Phil loved the Jewish traditions, but his problems learning Hebrew only added to his shame. He could not bear to think that he was letting his whole community down.

Given all of his problems dealing with printed texts, why did Phil imagine that he could ever become a writer? The answer was simple. Phil loved stories, and he loved the idea of sharing stories with others.

Telling stories was a tradition in Phil's family. No one was a professional writer, but Phil's father and uncles were natural storytellers. They always tried to outdo one another with their stories at family occasions, and Phil grew up with a love for the art of storytelling.

Phil often told his own stories to his mother and grandmother as they worked in their little house. Although he often had trouble remembering his school lessons, Phil had a fantastic memory for things he experienced and felt. Now he wove these memories into stories. As he told his imaginative tales, Phil could see in their eyes that his mother and grandmother saw him as something more than the failing student that his report cards, teachers, and tutor described. To them he was a spinner of dreams and a creator of worlds. Through their love and esteem Phil came to see himself that way, too.

As he grew older, Phil's schoolwork slowly improved. By the time he reached high school, Phil read well enough to pass most of his classes with above-average grades. He excelled in art and became the cartoonist for the school's literary magazine.

In his second year in high school, something amazing happened. For the first time Phil became so engrossed in reading a book that he read on and on without even realizing he was reading. The growth and development of his brain had finally reached the stage where it was able to support the complex functions required for reading without the correct kinds of instruction Phil had missed earlier, and his emotions, imagination, and even his senses became fully engaged. The experience was exhilarating. Phil felt like the narrator was a real person, the friend Phil had been looking for all of his life.

Phil repeated this experience again and again with other books. He absorbed the personal and private worlds of the other authors, taking in their perspectives and personalities. Phil realized that through his imagination he could see the world through others' eyes. Even better, using his imagination he could become another person entirely and imagine the world from a point of view totally different from his own. This is called persona writing.

Phil was now more certain than ever that he wanted to become a writer. He began writing short stories and poems for his school's magazine.

After high school Phil studied writing in college and then graduate school. He began to teach writing to college students. He showed them how to adopt a different persona in their imagination—personas with agendas and attitudes different from their own—and then to write from that perspective. Phil eventually founded his own school, the Writers Studio, where he has continued to teach these methods.

At the same time, Phil began trying to write professionally, and his career as a writer soon took off. He published a book of poems, and it was nominated for the National Book Award for Poetry. Although it did not win, this was a tremendous honor for the young poet, and a clear sign that Phil was truly a writer, just as he had always dreamed.

In the years that followed, Phil published many more collections of poems and became recognized as one of America's best poets. In 2008 he was awarded the Pulitzer Prize for Poetry, the highest award available to an American poet, for his book *Failure*.

In 2011, Phil opened a new chapter in the story of his life by publishing *My Dyslexia*, in which he talks about his own journey as a dyslexic person. Through this book Phil has become an inspiration and a friend to countless other dyslexic people. Although he still finds the act of reading words to be difficult, he knows that he has a mind that was quite literally made for stories.

Perhaps his most amazing achievement as a poet is his book-length novel in verse, *The Wherewithal*. In this book, Phil uses his gift for seeing the world through others' eyes to amazing effect. In telling his story through the eyes of a young Polish Catholic man working in the basement of a welfare building in San Francisco in the late sixties, he weaves fifteen different narratives into a deeply moving meditation on the Holocaust and the sufferings of the poor.

In his new memoir, *Comforts of the Abyss*, Phil describes how persona writing helped him overcome his often-paralyzing inhibitions of self-doubt and fear and turn them into creative inspiration. He tells us how other writers and mentors—such as Norman Mailer, Elizabeth Bishop, Joan Didion, and John Cheever—inspired him to turn his early struggles with dyslexia and immigrant striving into a way of helping others to become creative.

Phil spoke to us once about his own creative approach in words that mirror many of the things we've talked about in our discussions of creativity and memory. He said, "Dyslexia lends itself to original thinking, not rote formulas, because you can't do the formulas. You think up your own method based on intuition and instincts. Creativity is trial and error, trying to figure out a way to do something emotionally and intuitively." This is a perfect expression of how N-strengths use experience for creative purposes.

Listening to Larry Banks as he speaks confidently from the office he has occupied for the last decade as chair of the Media Arts Department at Long Island University, you might almost think he was talking about someone else's childhood when he described his early years in school.

"I first discovered I was dyslexic when I was about seven years old. It was clear that I was verbally bright, but when it came to reading, I was way below average. To stand in front of a class and read, which was standard in those days, was pretty much a nightmare for me. Finding out that I was dyslexic didn't really help, because I thought that was just another word for stupid. Everybody said I must have been stupid and possibly retarded, which was the word they used back then, and it didn't help that I went to a school where the only other Black kid besides myself really did have a severe mental impairment. So *retarded* seemed to fit because I couldn't do any of the things that the other kids were doing. That was my experience at the time. My mother was a teacher, and she was determined that she would find some way to get me past this and figure out what it was. So I went through a lot of remedial education, but for the most part it had absolutely no effect. School became just a series of hacks for me, learning to get by in my own way."

It was outside of school that Larry's real education took place. "I was that dyslexic kid who took stuff apart. My parents knew not to leave too many things around because I would take them apart. I wanted to understand what made things tick and how things worked and what they became. I was also especially fascinated by light. I'd wonder, what is light? . . . All the space around us is filled with light, but nobody sees *it*. Instead, I can see you because light bounces off you. That idea was fascinating to me. It still fascinates me. So these kinds of ideas were running through my mind as a kid."

For a kid who was interested in light and mechanics, Larry chose his family well: "My father had a lighting rental company for motion pictures, and from the time I was very, very young he used to take me around with him, so I basically grew up on the sets of motion pictures. Another saving grace for me was that when I was eleven, I discovered my grandfather had stored in his garage the equipment for a darkroom. So I set up a darkroom in the basement of our house, and I figured out how to develop my own film. I spent hours and hours playing around with the time and the materials and filters and seeing how these different effects changed the outcome. I was totally self-taught, and photography became a really strong skill for me. By the time I was fifteen, I got a job working at a camp as the camp photographer. As I got older, I also had access to lights and all kinds of equipment from my father to work with."

After graduating from high school, Larry went off to college, dual-majoring in communications and education. After college he felt drawn back to the film industry, and the contacts he'd made in the years he spent shadowing his father helped him land jobs as an electrician on a steady stream of film projects. Because of his deep familiarity with lights, Larry was soon able to transition to working as a gaffer, a chief lighting technician. "The gaffer is the hands and feet for the director of photography, so I became the one who put up all the lights. And I just had a kind of sense for how I should do that to get the right effects. I think that was part of the gift that dyslexia has brought me. My brain isn't mathematical. I didn't calculate that stuff. It was just like an extension of me. I could just picture in my mind what I should do, and I did that work for about ten years. I got a chance to work with some very impressive people and I learned a lot; but after a while the other members of the crew would say, 'You're really running this anyway, so why aren't you a DP?'"

Larry eventually started landing jobs as a director of photography, and in addition to several highly regarded feature films, he made countless commercials and music videos. Working as a DP was wonderful, but it raised a big problem. To prepare the lighting and photographic effects that were needed, Larry had to read the scripts. But he couldn't read the scripts—at least not fast enough to prepare for interviews or keep pace with filming.

Fortunately, Larry discovered the Kurzweil Personal Reader, which was the first text-to-speech device. It was literally a career-saver for Larry. As the Kurzweil reader read the scripts aloud to him, Larry could envision in his mind the particular look that would fit the scene—both the lighting and the camera angles and effects—and he would know what he needed to do to achieve that look. He could simulate it all in his mind.

Larry might have continued full-time as a highly successful cinematographer if his friend Spike Lee hadn't asked him to present some public lectures for LIU Brooklyn and FVA (Film/Video Arts) for young people interested in film careers. Larry's talks were very well received, and they soon led to many other talks and workshops. Larry found that he really loved sharing what he knew, and it also helped him to clarify his thinking about his own work. "Sharing forced me to do something that I hadn't done before, which is think about how I could break down what I do into a system that other people could grasp."

Looking into his own mind and reflecting on how he made his decisions about lighting, Larry realized that he had created a mental matrix of light characteristics. This matrix, or mental model, consisted of three dimensions, each with two poles: bright and dark; soft and hard; and warm and cool. He saw that he routinely used this matrix to characterize all the possible lighting effects he was trying to achieve. This made it easy for him to decide what lights he actually needed to use to achieve those effects. "I'd read a script and say, 'This is a five-warm, and a three-soft, and a four-dark.' Then for me to re-create that, it's pretty simple." Better yet, Larry found that this way of thinking about light was teachable, and that it could help his students learn how to set up lighting reliably to achieve certain effects, as well.

"Eventually some professors at one of my workshops came

up to me and said, 'Would you be interested in teaching at our school? You're very good at making sense of this stuff.' And I said, 'Yes, I'd like that very much.' But there was just one problem. To teach at the university level I'd have to go back to school and get a master's degree. That was a strange thing for me, because school was never my favorite place. But I felt like I almost had to do it, and I'm glad I did, because now I'm a professor who teaches film and digital production. My expertise is cinematography, and I love directing and producing, and I love working with students, and I love working with film.

"I've also been pleasantly surprised by the ways my ability to look conceptually at situations and at groups of people from multiple perspectives has served me as well in academia as it does when working on a production team. I can spot potential problems and areas where things will go off in advance, and hopefully head them off. One of the reasons I've been department chair for over a decade is that I understand all the different personalities and perspectives of my faculty and of the school leadership, and I'm often sought out by other professors for my insights. In a world of PhDs, it's sort of ironic, but the dyslexic person is the one they come to when they want to get a sense of the future. I can see the patterns and how they'll come out. To me, the outcomes are just obvious."

In reflecting on where his life has taken him and how different his view of his dyslexic mind has become since the days when he thought dyslexia meant stupid, Larry offered the following thoughts. "I always felt that there was something really positive that I got from dyslexia. I knew that I had the capacity to do certain things really well, but I had no idea whether that was because of or in spite of my dyslexia. Then I read the book *The Dyslexic Advantage*, and then all the pieces started to fall into place. I could look back on my life and say, 'That's why this' and

'That's where I came to this.' I saw that all these things were connected because of my dyslexia. It helped me rewrite over sixty years of my history, and now at this point in my life I can say, 'Because I'm dyslexic, I am able to use light to weave a narrative through images and pictures.'"

Part 6

D-Strengths

Dynamic Reasoning

25

The "D" Strengths in MIND

D-Strengths in Action: Sarah Andrews

When Sarah Andrews was a child struggling in school, her mother called her "my little underachiever." As you'll see, it's been a long time since anyone has called Sarah an underachiever.

Like many individuals with dyslexia, Sarah was a late bloomer. And like many dyslexic individuals, Sarah's "blooming" was less like the gradual unfolding of a bud into a blossom than the astonishing transformation of a caterpillar into a butterfly.

Sarah was born into an academically accomplished family: her parents were teachers and her siblings honor students. But right from the start she struggled with many academic skills, including spelling, math calculations and procedures (especially showing her work), and rote memory of all kinds. However, her biggest challenge was learning to read. Sarah described some of her early reading problems this way: "The letters and print vibrated on the page. I saw the fibers in the wood pulp in the paper. I put my head so close to the page that my chin was right on it, and my teacher put a ruler under my chin to keep my head off the page."

Because of her difficulties, Sarah was unable to get through a first-grade reader until third or fourth grade. She had to read each sentence several times to understand it, and she had difficulty keeping her mind from wandering, because each word "set off cascades of ideas and associations" that she felt she had to "test and integrate."

As Sarah recalls of her early years of school, "I was taking in tremendous amounts of information and starting to group it and sort it and arrange it, but there wasn't anything coming out—there wasn't a product." Sarah's mother—who also happened to be the English teacher at the small private school Sarah attended—was determined to fix this lack of output, so she drilled Sarah on writing, particularly on writing essays. To the delight of both, Sarah found that it was "easier to code than to decode."

Unfortunately, reading remained a problem. When Sarah reached high school, she still "couldn't read a lick," and it finally caught up with her on the SAT. Although she'd received "decent" grades in school, her performance on the SAT was so poor that it astonished her teachers. As Sarah recalls, "Someone finally asked if I finished the test. . . . I said, 'I got about halfway through.'" The riddle of her poor performance was solved—almost.

Sarah was packed off to the remedial reading lab. Because she could pass all the phonics tests but just couldn't read fluently or retain what she'd read, the reading teacher improperly stated, "You don't have dyslexia. You're just lazy." Despite this misdiagnosis Sarah practiced diligently, and she improved her reading speed enough so that when she retook the SAT, she not only finished but doubled her score. Even with this improvement, though, Sarah still hadn't reached grade-level reading proficiency. She was, in her own words, "a bright high school senior now reading at the eighth-grade level." To this day her reading remains agonizingly slow.

After graduation, Sarah turned down admission to two top art schools and enrolled at Colorado College. She wasn't sure at first what she wanted to study, and like many college students with dyslexia, she struggled in the courses she was required to take during the first two years. She took a course in poetry to meet her English requirement—primarily because she thought it wouldn't require much reading—and received the first and only F of her career. To make up for those credits, Sarah took a course in creative writing. To her delight and surprise, she found that she not only enjoyed writing fiction but actually had a knack for storytelling. Although she couldn't "sustain the activity" of her stories for more than three or four pages, her teacher praised her stories as outstanding miniatures. Sarah didn't realize it then, but this newfound skill would play an important role in her life.

Another life-changing discovery came when Sarah took her required science course. It was geology, which she selected almost on a whim because the only scientist in her family, her aunt Lysbeth, was a geologist. Also, like Sarah, Lysbeth was dyslexic.

Sarah soon realized that geology was "the right playground for my mind. . . . I at last found teachers who perceived my talents, and I could learn from maps and illustrations rather than insurmountable texts. For the first time I was around a concentration of people who thought like I did, and I wasn't being snubbed as a weirdo."

Even among these like minds, Sarah was delighted to find that some of her talents still stood out as exceptional. "I was the best map interpreter—I was really quick at taking in graphical information holistically, seeing the patterns, understanding their meaning, and making interpretations from them. Being the best at something in class was a new experience for me, so I stuck with it." As she did, the self-doubt produced by the earlier labels of "lazy" and "underachiever" began to fade away. "By the time I noticed that I was not in fact lazy, I had earned both a B.A. and M.S. in geology."

For her first job as a geologist, Sarah went to work as a research scientist for the U.S. Geological Survey. She was assigned to study sand dunes to help learn how gases and fluids could be removed from rocks that had been formed from similar dunes in prehistoric times. Sarah found that she was especially good at visualizing physical bodies three-dimensionally and at imagining how processes would act on those bodies over time. These skills made her especially good at detecting analogies between modern dunes and ancient rocks, and at predicting the structure and behavior of buried rock formations.

After leaving the USGS, Sarah went on to work for several oil and gas companies as an exploitation geologist (that is to say, a geologist who specializes in finding ways to remove known oil deposits from the ground). Sarah's role with these companies was to improve oil and gas extraction from drilled wells by predicting how these substances would move through the surrounding rocks. Here, too, her spatial imagery and pattern-reading abilities proved invaluable. Sarah was especially good at reading the wireline log—an immensely helpful but bafflingly complex visual readout of the physical characteristics of the rocks and fluids surrounding a well shaft. Sarah quickly learned how these squiggles on a page, which resemble the EEGs neurologists use to analyze brain activity, could predict the properties of surrounding fluids and rocks. Sarah found that she could transform these squiggles into mental 3-D images: "I could visualize in time and space how the oil was going to move through the rock, even when it was fragmented and shattered."

Clearly, Sarah's job seemed tailor-made to fit her mind. Geology drew heavily upon her strengths, placed little strain upon her weaknesses, and provided her with an endless string of fascinating puzzles to captivate her intellect.

The question we'll consider in the chapters ahead is: What *are* the strengths that so powerfully equipped Sarah for her work in geology?

Dynamic Reasoning: The Power of Prediction

Sarah's geological reasoning skills are no doubt due in part to her outstanding M-strengths. Sarah's powerful 3-D imagery system allows her to mentally visualize and manipulate lifelike full-color imagery, which she finds incredibly useful for tasks like map reading, navigation, and remembering 3-D environments.

Yet if we look closely at the full range of the reasoning skills Sarah uses as a geologist, we can see that they're not entirely spatial. Geological reasoning requires more than simply visualizing and manipulating images on spatial principles alone. It also requires the ability to *imagine* or *predict* how those images will change in response to *processes* that aren't entirely spatial in character, like erosion, earthquakes, sedimentation, and glaciation. These processes involve complex, dynamic, and variable blends of factors, and are themselves often subject to larger processes, like climate variation or plate tectonics.

We call the reasoning skills that are needed to think well about such complex, variable, and dynamic systems *dynamic reasoning*, or the D-strengths in MIND.

> D-strengths help accurately predict unwitnessed past or future states and solve complex problems by using episodic simulation default mode network mechanisms. D-strengths are especially valuable for thinking about past or future problems whose components are variable, incompletely known, or ambiguous, and for making practical or best-fit predictions or working hypotheses in settings where precise answers aren't possible.

From one perspective, D-strengths can be seen as a subset of N-strengths because they're based primarily on episodic simulation, which we discussed in detail in the last section. Yet D-strengths are important enough, complex enough, and distinctive enough in their applications that we believe they deserve their own listing within the MIND strengths.

The key difference between N-strengths and D-strengths is the distinction between *simulation* in general and *simulation for prediction or problem-solving.* As we described in chapter 19, N-strengths include all the functions of the episodic construction system, each of which works by combining elements of past experience to construct lifelike narratives or scenes. These scenes may restage actual past experiences—in which case we call them *episodic memories*—or they may recombine elements of past experience in entirely new ways to form *creative simulations* or *works of imagination.* Only when these creative constructions aim at predicting future events, reconstructing past events that we didn't witness, or solving new problems do we say that they involve D-strengths. When the episodic simulation system is used to recombine elements of memory to entertain, persuade, paint an arresting picture, or form a compelling vision—but not to predict or reconstruct actual events or conditions or to solve specific problems with precise constraints—we refer to those actions as N-strengths, not D-strengths.

D-strengths aren't simply creative in a general or unconstrained way. They aim to predict real states or outcomes—to simulate the world as it actually exists, has existed, or will exist. D-strengths use episodic simulation to construct *true narratives*. They are N-strengths in their most practical mood, with their work clothes on and their sleeves rolled up. N-strengths may simply aim at being compelling. D-strengths must be *accurate*.

Let's compare the similarities and differences between D-strengths and N-strengths by seeing how Sarah employs them in her two very different—yet surprisingly similar—careers.

26

The Advantages of D-Strengths

Dynamic Reasoning: Narrative + Prediction

The first time we spoke with Sarah, she quickly informed us: "I'm going to tell you everything as a story, because that's how I experience the world." The conversation that followed proved her assertion to be entirely true.

Similar to many individuals with dyslexia, Sarah has a highly narrative or episodic memory and reasoning style. She is a classic "family historian with a poor memory." As she told us, "I'm the family elephant: All my cousins come to me when they want to find out what happened when and where." Sarah also struggles with rote memory: "I have no rote memory at all. I can remember things only if they fit into a structure." Typically, this structure is a story: "Stories are what I remember—they stick in my memory."

In fact, narrative is more than a reasoning style for Sarah. It is also a second career. Sarah Andrews is the author of twelve highly regarded mystery novels that feature the exploits of professional geologist and amateur detective Em Hansen. In these novels, Em uses her skills as a geologist and her prodigious powers of episodic simulation to solve mysteries.

However, the first problem Em solved was one of Sarah's. Sarah told us, "When I was in my thirties and working in the high-stress atmosphere of the oil business, I had trouble settling down to do my geology work if I'd witnessed an event that had a strong story to it. But I discovered that if I went ahead and wrote down the story or anecdote . . . I could squeegee it out of my mind and focus on the work I was supposed to be doing. Writing just seems to get the stories out of me."

Sarah described the mental mechanism she uses to construct her stories in a way that closely mirrors the description of episodic construction: "As time passed, the anecdotes gathered like lint in my mind, so I made fabric from it, and the bits of fabric needed to be rearranged in order to move tensions and troubles toward resolution. . . . [This fabric formed] a patchwork quilt of memories, in which I took various events and reorganized them into new events."

Sarah soon realized that before she could turn these "patchwork quilts" into novels, she needed to figure out how to explain to general readers what geologists do and how they think. However, before she could do that, she had to explain those things to herself. This preparation required two solid years of introspection, but it was well worth the effort. In the end it enabled Sarah to describe geology in all its complexity and wonder in a way that her readers could understand. It also led to many deep insights about geological reasoning, which Sarah has described in several fascinating essays. One of these insights concerns the narrative nature of geology. Although the rocks and minerals on which geology focuses may initially seem nonnarrative and impersonal, Sarah insists that geology involves narrative every bit as much as mystery writing does because the rocks tell stories if we are trained to read them.

Sarah isn't alone in this view. Jack Horner, the world-famous paleontologist whom we met in chapter 14, has also identified a narrative element in geology. He describes it in his book *Dinosaurs Under the Big Sky*: "Geology is the most important science for a dinosaur paleontologist to know because we find dinosaur skeletons in rocks. Our knowledge of geology helps us understand where to look, what to look for, and how old the fossils are. . . . It helps paleontologists figure out what happened to animals, what may have killed the animals, and what happened to their remains after they died. Geology tells us the stories in the rocks."

Narrative construction in geology differs from writing novels largely in the constraints that reality places upon the construction process. While the novelist recombines experience to create an interesting and compelling scene that resembles something that *could* happen, the geologist uses observable facts to *predict* exactly what the earth's actual past conditions were or what future conditions will be.

Dynamic Reasoning: How Elephants Become Prophets

The role that episodic simulation plays in geological reasoning can be seen in the description that Sarah gave us of her research on rock formation: "I took in everything I had ever observed and projected myself backward in time, seeing the landscape on which the sands had been deposited before they became rock."

This description reflects several of the features of episodic construction that we've already discussed. *Taking in everything* means forming memories through experience and observation, so parts of these memories can later be used for episodic construction. *Projecting myself backward in time* means combining memory fragments through episodic simulation into mental scenes that predict what the past was like.

Sarah further described this process of construction when she wrote of thinkers like herself: "We are great sponges for *observed patterns*, both the concrete patterns of visual observations and the more abstract patterns of process and response. . . . Repeated patterns become ideas, and new patterns lead to new paradigms. . . . [We] can, using the barest shreds, 'see' through solid rock, back through time, and into future events." Importantly, this constructive process isn't just limited to building static or snapshot scenes of single points in the past or in the future. It can also create a continuous and interconnected series of scenes that allow the observer to "travel through time" in a form very much like a time-lapse film.

This ability to create *a continuously connected series of mental scenes* is especially valuable for imagining and predicting the effects of *processes* that take place over long periods of time, like erosion, flooding, or the movement of the earth's crust along fault lines. Each of these processes occurs at its own unique rate, or along its own time dimension. Episodic simulation can be used to predict these processes' combined effects without losing sight of the impact each exerts. This allows geologists to mentally experiment by independently varying these dimensions, changing the rate or extent of their effects. That's why it's the ideal tool for reading stories in complex fields and for generating hypotheses, evaluating solutions or plans, or predicting the possible results of different actions.

Sarah points out why episodic simulation is especially valuable in situations that are changing or ambiguous: "By working qualitatively we can mentally bridge gaps without having to plug in assumptions, and, as a result, it becomes possible to work with uncertainties, rather than simply overriding them." By *working qualitatively,* Sarah means working with episodic memory fragments that have been acquired from the original observations and can be used to construct scenes that simulate past, present, or future conditions in a form similar to the original observations.

We can get an idea of why this qualitative approach based on episodic simulation is so powerful by comparing it with its alternative: *abstract reasoning.* Abstract reasoning uses verbal, mathematical, or symbolic abstractions rather than episodic memory fragments. These abstractions have been created by *combining and converting* the original observations into generalizations that are stored in semantic memory as decontextualized facts. These abstract generalizations are often useful for reasoning about routine or typical cases, but less useful for thinking about unusual, unexpected, or unprecedented cases.

We can highlight some of the benefits of using more primary or separated data with several examples. Think first of a baseball team with thirty players. The team's batting average can be calculated by combining the personal averages of all the players. Let's say our team's average is .250, or one hit in every four at bats. Now, most of the individual players will have averages that are close to .250, but a few poor hitters may have averages around .100, and a few excellent ones may have averages near .350. If we want to predict future performance, the team average will be fairly good at predicting the batting success of the team as a whole. It will also be fairly good at predicting the success of the average hitters. But this generalized average will be poor at predicting the performances of the best and the worst hitters. We could much more accurately predict the performances of these outliers by using their past *individual* performances.

Here's a second example that also demonstrates how qualitative reasoning can be better in new or unprecedented situations. Let's say that we want to predict how a particular batter will perform against a pitcher whom he or she has never faced. That player's overall batting average—which combines the results of batting against all pitchers—will be less useful than considering how that player has batted against pitchers with styles similar to the new pitcher.

These examples should help to illustrate some of the advantages that D-strengths can have when working in situations where the important information is changing, incompletely known, or ambiguous. By building mental scenes using observations from the real world rather than verbal or numerical abstractions, we can arrive at practical, best-fit solutions for difficult, unusual, or unprecedented problems without having to assume away ambiguities. This process of using what fits rather than relying entirely on abstract analysis or secondhand models is extremely powerful for solving practical problems.

27

Trade-offs with D-Strengths

Nonlinear Reasoning: Mental Simulation, Mind-Wandering, and Insight

While D-strengths can provide tremendous advantages in dynamic or changing situations, they can also create drawbacks in other settings. One of the biggest drawbacks is a reduction in speed and efficiency and a need to take some time for reflection.

When we ask dyslexic individuals with strong dynamic or narrative reasoning styles to describe their thinking process to us, many describe a kind of backward process in which the answers come first, essentially fully formed, and usually after a period of calm reflection or mind-wandering. Then, only after the answer has arrived essentially fully formed does a more conscious analytical process begin, in order to discover *how* this answer connects with the initial conditions.

Douglas Merrill's description of his problem-solving method is typical: "I usually begin by visualizing what I think the end stage should be, then I work backward. I can't exactly describe what I do because it feels more intuitive than traditional storytelling or deduction."

Sarah Andrews echoed this description when she wrote of herself and others like her: "Given a problem and an hour to solve it, we typically spend the first three minutes intuiting the answer, then spend the other fifty-seven backtracking . . . to check our results through data collection and deductive logic." According to Sarah, this intuitive approach "functions in leaps rather than by neatly ratcheting intervals" and is "less lineal than iterative or circular."

This intuitive approach is used very heavily by individuals with dyslexia who excel in dynamic and narrative reasoning. It can be very powerful, but it does present a problem: when viewed by those on the outside, it can look an awful lot like goofing off. Sarah shared an example of this from her own life. One day at work she was standing by her office window staring serenely out at the mountains while trying to let her mind "ease itself around a problem." Her CFO walked by her door and looked in. All he could see was one of "his people" staring out the window, so he snapped at Sarah to get back to work. Sarah calmly replied, "You work in your way, I'll work in mine. Now stop interrupting me." Sarah later wrote of this episode: "What this CFO didn't know was that staring into space is precisely how we work. It is our capacity to throw our brains into neutral and let connections assemble . . . that makes it possible for us to see connections that others can't. We relax into the work."

This need for patient reflection can also create enormous problems at school, where time for reflection is in critically short supply. Try convincing a teacher that staring out the window is how you do your best work, or that if you "get busy," you'll get less done. Yet this reflective approach really is a valid problem-solving method, and there's plenty of scientific evidence to support its effectiveness. In the research literature, this method of problem-solving is referred to as *insight*, and as we'll see, it's one of the important functions supported by the default mode network.

Insight involves the sudden recognition that connections exist between elements of a problem. The classic historical example of insight is Archimedes shouting, "Eureka!" and leaping from his bathtub as he suddenly realized how water displacement could be used to measure the volume of irregular-shaped objects.

Insight is at its most useful when step-by-step or analytical problem-solving is hampered by the ambiguity or incompleteness of the information—that is, in situations where D-strengths are required. Insight is also highly dependent upon the I-strengths we described in part 4. This is because insight depends on the ability to make the same broad and distant *cognitive* connections between concepts and ideas that are characteristic of I-strengths. Because insight is so closely linked with D-strengths and I-strengths, one would expect to find that individuals with dyslexia are especially good at insight-based problem-solving. In our experience, that's precisely what we do find.

Although insight-based problem-solving is very powerful, much of its connection-making process takes place outside the person's conscious awareness. That can make it seem mystical, shoddy, or even slightly disreputable. But there's an observable neurological mechanism underlying insight that's been well worked out over the last decade by researchers.

One of the scientists who has contributed most to our understanding of insight is Dr. Mark Beeman, whom we first encountered in chapter 4 when we discussed his work on our brain hemispheres' different approaches to language. Mark's work has been especially important in describing the different phases that make up the process of insight.

In the first phase, the mind focuses actively upon the problem at hand and sets out the questions that need to be answered. This highly focused phase quickly gives way to a *relaxation* phase, where the mind loosens its focus and begins to wander. As Mark has described, "There's an overall quieting of the brain's processing, because it's trying to calm everything down and wait for something to pop out." That "something" the brain is waiting on is the recognition of distant or novel associations, which are just the kinds of connections that individuals with dyslexia typically excel at making. When a suitable connection is found, it results in the simultaneous activation of a broad cellular network that stretches all over the brain. This widespread electrical burst creates the subjective sensation of the eureka moment.

Notice how closely this insight mechanism links mental states like relaxation, reflection, and daydreaming with productive abilities like creativity, the ability to detect distant connections, and the ability to solve problems. Not surprisingly, the brain circuits that become activated during daydreaming or mind-wandering include the brain's default mode network and episodic construction system, which we described in our discussion of N-strengths. In other words, daydreaming consists of free-form and undirected scene construction, or the creative recombination of episodic memories. This is largely what *imagination* means. Small wonder there's such an extensive overlap between imagination and insight, daydreaming and problem-solving—or between staring out one's window at the mountains and solving difficult geological problems.

Factors like emotional well-being and positive mood also seem to play an especially large role in supporting successful insights, and they work by enhancing the relaxation phase. This is why many great insights seem to occur when you are showering, bathing, sleeping, wandering the beach, gazing out windows, or staring into space. Attempting to "force" or hurry insight only inhibits it. This is one of insight's most confusing features: its success seems to vary almost inversely with effort, so the deepest engagement requires a kind of deep disengagement. The harder you try to solve some problem using insight, the less likely you are to succeed.

This isn't the first time in this book that we've stated that tight mental focus and attention can inhibit creative connections. Think back to our discussion in chapter 15 of latent inhibition, where we mentioned that tight mental focus and resistance to distraction are inversely correlated with creative achievement. In contrast, making distant, creative, insightful connections may be fostered by a slightly leaky attentional system, which allows ideas to mix.

Childhood may be the time in life above all others when nature favors us with the capacity to make creative and insightful connections. Mark Beeman speculated to us on the value that the human species' unusually prolonged period of attentional immaturity may play in the development of creativity: "Perhaps there's some benefit in the delayed development of mental focus—maybe that's why humans in general develop so slowly. And perhaps those children who are developing more slowly in their attentional skills are developing more richly in certain aspects of creativity; and perhaps this extra creativity means that they'll actually develop in great ways if we don't mess them up too much in the meantime."

Mark was clear about the kind of "messing up" he had in mind. "One big concern I have is with the use of stimulant medications used for ADHD [e.g., Ritalin, Concerta, Adderall, Vyvanse].

These drugs improve mental focus and resistance to distraction, but getting people to focus more may ultimately be bad for creative thinking. We may actually be inhibiting growth in areas like creativity and insight that are very useful—and that's something that we really should not do unless it seems absolutely necessary."

We agree wholeheartedly. Rather than judging a child's development solely on qualities like speed, quantity, and focus during work, we should be monitoring their development of creativity, use of insight, and time spent in reflection. By failing to recognize the value of the slower but incredibly rich insight system, and instead placing all our emphasis on linear, rule-based, deductive thinking styles, we hinder the development of all children, but perhaps especially those who are the most creative and insight dependent. This does not mean we shouldn't help children develop better mental focus and organizational skills; but it does mean we need to do a better job of recognizing the value of these different attentional set points, and learn to work with them in developing their unique potential, rather than try to deform them into something they were never intended to be.

One area where we often see insight-based problem-solvers suffer unnecessarily is in math. We frequently met phenomenal young mathematicians in our clinic who were unable to show—or in some cases even describe—the steps in their work, yet who got nearly every problem right. These students were solving problems through insight and mental simulation—matching the patterns of new problems with ones they'd seen before, and sorting through their memory stores for answers that fit rather than employing step-by-step analytical reasoning. While it's important to help these students learn to trace through the intervening steps, in the preteen years, so long as they can consistently reach right answers, they should be given credit for their understanding even if they fail to demonstrate all the steps in their work. As they age and their neural systems mature, their efficiency at moving between insight and analytical problem-solving will improve, and they'll be better able to reverse engineer the intervening steps and show their work. Unfortunately, we've seen some truly disastrous cases where profoundly gifted young mathematicians have lost their love for math simply because too much emphasis was placed on having them show their work when they weren't developmentally ready to do so.

It's important to recognize that certain individuals are predisposed to solving problems through insight and mental simulation rather than through analysis. In our experience, this is true of many individuals with dyslexia. People whose reasoning is based primarily on these approaches can sometimes look unfocused, inefficient, nonlinear, or slow to others who don't fully grasp the nature of the mechanisms they're using, and they often have difficulty getting others to accept the results of their reasoning processes if they can't show their work. However, these methods of reasoning deserve far more respect than they generally receive. As teachers, parents, coworkers, and bosses, we need to be watchful for individuals who frequently reach the right results through rather opaque methods, and when we find them, we need to treat their different reasoning styles with the seriousness they deserve. Not all staring out the window is productive reasoning, but quite a lot is; and it's important to understand that some people—including many of the most creative—really do need to relax into their work.

28

D-Strengths in Business

D-Strengths in Action: Glenn Bailey

We've seen how dynamic reasoning can help in situations that are changing, uncertain, or ambiguous. Now let's look at some intriguing evidence that demonstrates how individuals with dyslexia have achieved conspicuous success in one of the most changeable and uncertain environments of all: the world of business.

We'll start by looking at one dyslexic entrepreneur who's shown a knack for building successful businesses. His name is Glenn Bailey.

Since Glenn began his first company at the age of seventeen, he's built many profitable businesses in a wide range of sectors including services, construction, and retail. Yet for all his success in the business world, Glenn rarely found success in school.

"My school career was dismal. I had a hyperactive mind, so my focus was just not there. My mind tended to wander a lot—usually in order to entertain myself. I'd be off in another world and thinking about other stuff.

"My mind is very visual: I can see anything in pictures, and I always visualize things. I can't help it. It's how I'm wired. So whatever you talk about, I'll see pictures in my head. Very vivid, colorful, lifelike pictures. They aren't still pictures. I can make them move. Reality, fiction, whatever. I really have to pull it back in to get focused. It was also a problem in the classroom because I'd sit there and imagine where I'd want to be and what I'd want to do and what I wanted to become, and I'd think happy thoughts, and I'd just be tuned out the whole time in class. I'd sit there nodding and smiling, but really I was like, *What are you talking about?*

"I was also very curious, and strangely enough, that became an issue at school. I'd ask questions like 'Why's this?' or 'Why's that?' and that was treated like a problem. Teaching was regarded as a one-way street, but I really learned best through interaction.

"My biggest problem was in English. Being able to read is the 'face of intelligence' you present to society, and if you can't read, people just automatically assume that you're stupid. What happens to individuals with dyslexia in school is that reading becomes this big fifty-pound weight that just drags your whole body under. So I didn't have much confidence in any department of academics. I just thought I wasn't that bright. I called myself 'the Shadow' because I was just trying to get by and go unnoticed day by day."

Glenn left school at age seventeen when he found it wasn't adding much to the strengths he intended to use in the real world. "The last day I ever went to school I found myself studying for English—which I didn't excel in or even like—during my math class. And I thought that was quite ironic, because I love math and I love numbers. They're very logical, and I can do math in my head quite quickly. Yet here I was, studying for a course that I hate—English—in a math class so that I wasn't learning math. So that was it for me. I left school, and that became a huge trigger point in my life. I said, 'Look, I can't do anything about the last sixteen to seventeen years, but I can do a lot about the future. Where do I want to be in the next five, ten, fifteen, twenty, twenty-five years?'"

So Glenn set goals, and he began pursuing them. After opening and running his own ski shop for several years, he spotted an opportunity to introduce bottled water to Vancouver. At the time he was twenty-three. Within ten years Glenn's company had 32,000 accounts and annual revenues of $14 million. He was so successful in building the Canadian Springs water company that he became known in his hometown of Vancouver as "the Water Boy." In 1996 Glenn sold his share of the business for $24 million and received a Business Development Bank of Canada Young Entrepreneur Award. Since then, he's built many more successful businesses, including another water business that provides on-site purification.

When we asked Glenn to identify the keys to his success as a serial entrepreneur, he mentioned his abilities to spot opportunities and develop a vision, but he also cited his ability to form relationships with other people: "For me—as for any dyslexic—it's about having the right people around you. Motivating and delegating is a massive part of what I do. You can't do everything by yourself. I rely on people a lot, and on their support. I also have amazing family—parents who always loved and believed in me absolutely—and friends, and teachers, and my wife is absolutely super. Everything I've ever done I owe to them."

Dyslexic Entrepreneurs: A True Growth Story

Glenn Bailey is far from alone in being a successful dyslexic entrepreneur. At the time of our first edition, a Google search on the term "dyslexic entrepreneur" called up over 37,000 links. Eleven years later, it calls up 6.37 *million*. Many of these links bring up lists of dyslexic entrepreneurs or biographies like Glenn's. But a large number also link to the work of Dr. Julie Logan, who was professor of entrepreneurship at the Bayes (formerly Cass) Business School, City University of London. Julie isn't dyslexic herself, but for over a decade she studied many dyslexic entrepreneurs and published several widely quoted studies on this research. We spoke with Julie and asked how she first became interested in the special abilities of dyslexic entrepreneurs.

"I used to spend a lot of my time doing management training and strategic development for managers in large firms. Then I moved over to working with entrepreneurs. I quickly noticed a difference. While many of the entrepreneurs were very good at presenting a clear and convincing picture of the business they were trying to build, they were very reluctant to get their ideas down on paper and to produce a written business plan. That was something I hadn't come across when I was working with corporate managers. A manager with a large company could typically produce a very good written strategic plan. So I found that rather strange, and I began to wonder why that was. That's really how I first got interested. After that, whenever I met entrepreneurs who were great at communicating visions but seemed reluctant to put their ideas down on paper, I started questioning them about how they'd done at school, and whether any of their children were dyslexic, and so forth. And that's how my interest in dyslexic entrepreneurs developed. That pattern kept coming up again and again."

Julie eventually conducted formal research on entrepreneurs, both in the United Kingdom and later in the United States. In the United Kingdom she found that the incidence of dyslexia among entrepreneurs was about twice the rate it was in the general population, and in the United States it was at least three times as high. In the United States, fully 35 percent of the entrepreneurs she surveyed were dyslexic, while fewer than 1 percent of U.S. corporate managers were.

Julie found several key traits among dyslexic entrepreneurs. The first is a remarkable sense of vision for their businesses. "They've got a very clear idea of where they're going and what they're doing, and holding that end point in sight is a very powerful tool because it can be used to harness other people around that vision." In other words, just like Gerry Rittenberg, whom we discussed in chapter 3, they've created a mental model in their mind and used it to simulate the future.

Second is a confident and persistent attitude. "Having got through and solved all the problems of schooling and coped with those challenges, they have a can-do approach that they bring to new situations. They not only know what they want to do, but they're also confident that it's going to work."

Third is the ability to ask for and engage the help of others. Julie found that most dyslexic entrepreneurs employed significantly larger staffs than nondyslexic entrepreneurs, and they were more likely to delegate operational tasks to their staffs while focusing their own attention on the overall vision and mission of their companies. "They know they're not particularly good at the details, so they surround themselves with people who are good at doing finance or good at attention to detail or whatever it happens to be; and unlike many entrepreneurs who won't delegate and will constantly interfere, they'll bring the best people around, and then they'll trust those people to do it. Many individuals with dyslexia have used the same sentence to describe this attitude to me: 'I employ the best people I can find, even if they're cleverer than me.' That comes out again and again."

The fourth trait is excellent oral communication, which they use to inspire their staffs. "Often they have very personal charismatic relationships with the people who work for them, and even when they have enormous empires, they manage to somehow create those same sorts of relationships. The employees are energized by them. A good example of this is how the staffs of both British Airways and Virgin Atlantic were both going to go on strike. They both wanted pay rises. But Richard Branson [the dyslexic CEO of Virgin Atlantic] went to the staff conference, spoke to the staff, and explained why he couldn't give them a pay rise; and everybody went back to work and carried on." By contrast, the CEO of British Airways didn't use that approach, and their labor standoff continued.

A fifth and final strength, which Julie has repeatedly observed but has not yet confirmed with her research, is that "many of these successful entrepreneurs use their intuition a lot. For example, I've been talking recently to a very successful dyslexic entrepreneur, and he told me that he never does formal market research. He just goes and he'll stand next to a store that he's thinking of purchasing, and he looks at the footfall [foot traffic] and that sort of thing. It's certainly more of a right-brain approach than a logical reasoning approach."

In short, the skills that Julie has identified in these dyslexic entrepreneurs consist largely of the kinds of D-strengths and N-strengths we've covered in this part and the preceding one:

dynamic reasoning to "read" future opportunities, establish end points, and solve problems, and narrative reasoning to convey the vision to others, inspire them, and persuade them to join in. These are the kinds of skills needed to operate in any environment that is changing, uncertain, or incompletely known.

29

Key Points About D-Strengths

In these chapters, we've discussed the D-strengths that many individuals with dyslexia exhibit. Key points to remember about dynamic reasoning include:

- Dynamic reasoning is the ability to read patterns in the real world that allow us to reconstruct past events we haven't witnessed, predict future events, or solve complex problems with defined parameters using episodic simulation.

- It is especially valuable in situations where all the relevant variables are incompletely known, changing, or ambiguous.

- Dynamic reasoning employs mental simulation and insight-based processing, which can yield valuable answers but whose steps can be hard to explain to others.

- Individuals with dyslexia who possess prominent D-strengths often thrive in precisely the kinds of rapidly changing and ambiguous settings that others find the most difficult and confusing.

D-Strengths in Action: Vince Flynn

In closing, let's look at an individual who showed remarkable D-strengths—novelist Vince Flynn. Prior to his tragic and all too early death of cancer at age forty-seven, Vince had written fourteen counterterrorism-themed novels that had sold over 20 million copies, making him one of the top-selling novelists of his era. Vince's stories were renowned for their intricate plotting and their surprising twists and turns, but none of his books exceeded his own life story in providing unexpected plot developments.

Vince was diagnosed with dyslexia in the second grade, when he was given the label SLBP (slow learning behavior problem) and placed in a special ed class. Vince recalled his problems with reading, writing, and spelling as "nearly incapacitating." He got by primarily by becoming a master of classroom survival tactics. As he once recalled, "[I] knew how to figure out the game and was always respectful to my teachers and tried hard. As long as you did that, it didn't matter how poorly you did on tests; they were going to pass you. . . . You participate in class discussions and the teacher says, 'This kid gets it; he just doesn't test well.'"

Vince graduated from high school with a C+ average, then enrolled at the University of St. Thomas in St. Paul, Minnesota. There his lack of reading and writing skills immediately began to cause problems. After his first semester he was placed on academic probation. He took two English classes and got a C− in both. He barely scraped by until his junior year, when two things happened that changed his life. The first was an event Vince described to us as one of the most humiliating of his life.

"I was taking a class pass/fail, and I handed in a paper to the teacher, but I wasn't there the next class to pick it up. Instead, I had a so-called friend pick up the paper, and later that day I went to meet him for lunch, and there was my paper sitting right in the middle of the table with eight guys gathered grinning around it. And at the top of the page there's a big red F, and toward the bottom the professor had written: 'I don't know how you ever got into college, and I don't know how you're ever going to graduate, but this is the worst paper I have ever read in all my years of teaching.' I was so embarrassed that I said to myself, *This is it, I can't keep going on like this. I've got to face this head-on.*

"Right after that, Al McGuire [the Hall of Fame college basketball coach and TV commentator] came through St. Thomas on a speaking tour, and I was thunderstruck by what he had to say. He talked about how he had grown up dyslexic, and he didn't know how to read or write; but he was a standout basketball player, and like me, he BS'd his way through high school and college with never better than a C average. After playing in the NBA for a few years, he began coaching the Marquette Warriors, and in his last season there they made it to the NCAA championship game. About fifteen minutes before tip-off the scorekeeper came up and handed McGuire the scorebook and said, 'Coach, I need your starting lineup.' And McGuire broke into a cold sweat because he had no idea how to spell his players' names. So he panicked and said, 'I can't—I have an emergency,' and he ran back to the locker room and locked himself in a stall, and he started praying, 'Dear God, if you let me win this game, I'll go back to night school and learn how to read and write.' So Marquette won, and Al McGuire retired from coaching and went back to school.

"Now, what I learned from this story was that the longer I put this off, the worse and more embarrassing it would become. So I really got serious because I wanted to write without being embarrassed and be something more than a functional illiterate.

"So I went out and bought two books, *Fundamentals of English* and *How to Spell Five Words a Day*, and I started going to the library every day and really working through them. I also started reading everything I could get my hands on, because I knew that was the only way to get better."

Vince started by reading *Trinity* by Leon Uris. Although he struggled during the first hundred pages, he was soon hooked, and he stopped worrying if he couldn't decipher all the words. He found that his mind could "fill in the blanks."

After graduating from college, Vince took several jobs in corporate sales while trying to enter Marine aviator school. However, the several concussions he'd suffered playing football disqualified him for flight training. He spent several years pursuing a medical waiver that would allow him to fly, but when he hit twenty-seven and was no longer eligible for officer training, he began searching for another direction for his life. "I remember saying to myself there's no way I'm going to spend the rest of my life sitting in a cubicle," so he started working on the manuscript that would become his first novel, *Term Limits*. Soon after, he "burned his boats," quitting his job and telling his friends and family he was going to make his living as a novelist.

We asked Vince what convinced him he could make a living as a novelist when only a few years before he'd been told that his writing set new standards for ineptitude. He responded that it was the confidence that came from finding that his mind worked a lot like the novelists he loved. He said that when he read or watched movies, "I always knew what was going to happen next. A lot of times I could tell almost from the first chapter how it was going to end." That experience made him think that if he could predict the plots of other people's novels, he could probably create his own.

Vince's suspicions were on solid ground. As we've discussed, prediction and narrative are closely related mental skills. However, Vince's ability to predict plot twists might not have been enough to make him risk his livelihood if he hadn't also seen other signs that he had special skill in dealing with complex patterns. One was a surprising talent he'd shown as a child: "I was a naturally gifted chess player. It was a weird, weird deal: even though I was failing in school, I could always just see in advance how the game would unfold. So I used to get driven around by my parents and dropped off at the houses of some of the best chess players in the Twin Cities. One year I actually finished fourth place in state. But I never told my friends at school because I was embarrassed about it."

Another hint came in college. "While I was struggling in my other classes, I ended up taking macroeconomics, and for the first time in my life I found myself sitting in a classroom being one of the only people who knew what was going on. A lot of the kids who were gifted in English and math were just scratching their heads because it didn't make sense to them—there were just too many variables—and suddenly I'm the one who's saying, 'How do you guys not understand this stuff? This is easy!' It just made sense to me."

Vince's powers of prediction were clearly revealed in his writing. After his first novel got him a contract with a major publisher, Vince chose a topic for his second novel that most people were largely unaware of in 1998, but which struck him as the most important national security issue of the time: Islamic radical fundamentalism. Vince's next three novels all focused on this threat, and all were published before the tragic events of 2001 occurred. "Prior to 9/11 I wondered, *Why isn't anyone else scared about this?* To me it was just obvious that this was a disaster waiting to happen."

Vince's ability to predict both the headlines and the behind-the-scenes action was so startling that one of his books, *Memorial Day*, attracted a security review by officials at the Department of Energy because they were certain he must have been fed classified information from an inside source. But Vince said the realism of his stories came entirely from his ability to predict using freely available information. "I can honestly tell you that I've never had an active-duty person with the CIA or the Secret Service or NSA or the FBI or armed services—or anywhere—give me classified information. I just take the information that's available in the public domain, and then I fill in the blanks. I have a way of connecting those dots."

People outside the government also noticed Vince's predictive powers, and some of them were so impressed that they were ready to stake money on them. "I was recently asked to go on the board of directors for a hedge fund. These guys said, 'We think that you have a good way of strategic thinking,' and they don't know why, and they don't know a thing about dyslexia, but they said, 'We'd really like to get you on this board so you can do some strategic thinking for us.'"

Vince also tried to convince others of the predictive power he believes many individuals with dyslexia possess. "I have a friend who's on the boards of several charitable foundations, and I'm always telling him, 'You know, every one of those associations needs at least one dyslexic on the board.' When he asks me why, I tell him, 'We just see patterns in advance. We could do you a lot of good.'"

Vince attributed this dyslexic predictive power to both nature and nurture. "Dyslexics are wired a little differently, and that probably makes us a little more creative innately, but there's also the impact of experience. School's like a wall, and for everybody else there's a ladder there, and they just get up on that ladder and climb over the wall. But for whatever reason, dyslexics don't know how to climb that ladder, so we've got to figure out another way to get past that wall. We've got to dig a hole under it or find a rope to build a rope ladder or find some other way around it. We're constantly trying to solve a problem, and I think that's one reason why so many dyslexics become inventors and creators—because they're constantly looking for ways to beat that system or improve that system or change it so it makes sense to them. This develops their skills in forecasting as well. So many things in life are like an algebra equation where you're given four knowns and three unknowns, and you've got to solve that problem. Dyslexics are forced to do that so much in their everyday lives that it helps us become really good at solving complex problems."

After listening to Vince list all the benefits he saw to being dyslexic, we ventured to ask him, "So you really don't find it surprising that we've titled our book *The Dyslexic Advantage*, do you?"

"Hear, hear!" he responded with a hearty laugh. "You're correct. I most certainly don't!"

30

What's New with D-Strengths

In this chapter we'll focus on new information that helps to clarify the nature of dyslexic D-strengths. As in our updates on the other strengths, we'll begin by looking at what dyslexic and nondyslexic individuals had to say when we asked them about their D-strengths.

What Dyslexic Adults Told Us About Their D-Strengths

The following are the questions we gave to our pool of dyslexic and nondyslexic adults on our MIND strengths surveys relating to D-strengths. Notice how they relate to the themes of mental simulation under real-world constraint, insight, and mind-wandering that were discussed in the previous chapters. As with the M-strengths, I-strengths, and N-strengths questions presented earlier, the numerical data indicating strength of agreement can be found in the Appendix. Here are the questions:

1. I am good at predicting how current events or processes will unfold and what is likely to happen next.

2. It often seems obvious to me what direction events will take in the future.

3. I often see several steps ahead of those around me.

4. When I watch a show or read a story, I can often predict what will happen next or how the plot will develop.

5. I often spot new and developing trends before others.

6. I am good at connecting the dots and spotting patterns or processes that others miss.

7. I am good at mentally simulating (or imagining) how a complex system or process will work, and how changes in conditions and components will affect their outcomes.

8. I am good at troubleshooting complex systems or processes and finding out why they're not working or where things went wrong.

9. I am good at developing creative strategies for reaching difficult goals or solving complex problems.

10. I often seem more comfortable than most other people in situations where there are a lot of unknowns or where things are changing rapidly.

11. I often seem to have a better sense of what to do than most other people in situations where there are a lot of unknowns or where things are changing rapidly.

12. I'm at my best in situations where there are a lot of unknowns or where things are changing rapidly.

13. My mind often solves problems or reaches the answers to complex questions through sudden insights, intuitions, or hunches.

14. When I solve math problems I often see the solutions suddenly, then have to go back to figure out the steps to reach them.

15. To answer a difficult problem, I often have to switch gears and take a break or work on something unrelated.

16. I spend a lot of time daydreaming and letting my mind wander.

17. Some of my best work is done when I'm daydreaming or letting my mind wander.

18. Other people often find my thinking nonlinear.

19. I often have difficulty explaining to other people how I come up with the answers I do.

Our dyslexic respondents resonated strongly with these D-strengths questions, with about 76 percent agreement plus strong agreement. In contrast, less than half as many of the nondyslexic respondents, 37 percent, showed similar levels of agreement.

Questions 1 through 9 deal with strengths including being able to predict or anticipate what's next; to spot trends or patterns; to mentally simulate those processes or trends; to use those mental simulations for troubleshooting; or to use them for formulating creative strategies. All these abilities deal in various ways with the core D-strength of using episodic simulation in situations where there are real-world constraints that limit the range of acceptable outcomes—that is, where accurate *prediction* becomes important.

Questions 10 through 12 demonstrate the usefulness of these experience-based simulation systems for dealing with uncertain or changing situations. Think back to our discussions in earlier chapters of how these systems, which heavily employ right-hemisphere processing, are used preferentially in situations that are especially novel or complex or that require consideration of extensive contexts to understand.

Questions 13 through 19 all relate to the rather mysterious form that problem-solving through insight, mental simulation, or the other operations linked to the default mode network can take when compared with more linear or deductive processes. Of all the D-strengths questions, many of these showed the greatest relative differences in the responses between dyslexic and non-dyslexic participants.

Together, these questions help to reveal the range, the power, and the often-confusing nature of the functions overseen by the systems that contribute to D-strengths. Let's look now at the research that's helped immensely both to better define these systems and to explain the nature of the dyslexic advantage.

New Research Relevant to D-Strengths

It has become clear over the last decade that dynamic reasoning's special powers arise from the actions and interaction of the default mode network, the episodic simulation system, and the hippocampal grid-cell systems. Over the past decade great progress has been made in our understanding of how these systems function and the skills they support. Although none of the research in this area has focused specifically on dyslexic minds, it has great relevance for them.

In previous chapters we've discussed how the hippocampal grid cell, episodic scene construction, and default mode network systems can work together to support many of the skills that are relevant to the MIND strengths, including spatial reasoning, concept formation, divergent creativity, episodic memory formation, scene construction, and imagination. These systems have also been shown to work cooperatively on many additional higher cognitive functions, including:

- introspection
- daydreaming and mind-wandering
- planning for the future
- forecasting or predicting future states or outcomes
- counterfactual reasoning
- perspective shifting
- storytelling and creative writing
- creation of visual arts
- complex problem-solving
- generation of vivid mental imagery
- and even reading

In the rest of this section, we would like to focus on one of these functions in particular: *prediction*. Prediction is in many respects the primary function of dynamic reasoning, and the mechanisms supporting prediction are now far more clearly understood than they were when our first edition was published.

Predictions: Big and Small

The first source of the predictive power of D-strengths is the ability to create accurate mental models of the physical and conceptual worlds using the hippocampal grid-cell and episodic construction systems. Using these models, we can run simulations where we manipulate the different parameters or dimensions of our models, and these simulations will allow us to see or predict what might happen given various changes, actions, or events.

We can divide the kinds of predictions we can make using our models into two main categories. The first category consists of what we'll call *big-P* Predictions. Big-P Predictions are the kinds of intentional, goal-directed predictions regarding possible future states or outcomes that we make with full conscious awareness. Many of the tasks that we listed in the previous section can actually be viewed as predictive processes that fall into this category. For example, problem-solving can involve mentally simulating potential solutions to predict which will have the best outcome. Planning and forecasting involve making predictions about future states, conditions, or events. Perspective shifting involves making predictions about what things might look like (either literally or metaphorically) from some other position. And so on down the list.

These kinds of conscious big-P Predictions are some of the most powerful D-strength skills, yet they comprise only a small part of the predictions actually made by these systems. During each moment of our waking lives, our brains are also constantly using these systems and the mental models they support to make moment-by-moment predictions about what we expect we'll experience next. Most of these *small-p predictions* remain hidden beneath the level of our conscious awareness, so we're not even aware of making them. The more accurate and efficient our models of the world become in predicting what comes next, the less often the outcomes of these predictions will rise to the level of our conscious awareness. Jeff Hawkins, whom we mentioned in earlier chapters, explains this process very neatly in his book *A Thousand Brains*:

> To make predictions, the brain has to learn what is normal—that is, what should be expected based on past experience. . . . [T]he [brain] learns a model of the world, and it makes predictions based on its model. . . . [It] continuously predicts what its inputs will be . . . and when [these] predictions are verified, that means the brain's model of the world is accurate. . . . We are not aware of the vast majority of these predictions unless the input to the brain does not match.

What happens when we experience a mismatch between our model's predictions and our actual experience? In Hawkins's words, "A mis-prediction causes you to *attend to the error* and update the model." In other words, mistakes grab your attention and set your mind to work to figure out and fix what went wrong.

The Surprising Value of Mistakes

The technical term for the mental state induced by mismatches between your unfolding experiences and your predictions is, unsurprisingly, the state of *surprise*. We'll describe what *surprise* means in terms of brain activity in just a moment, but at the subjective level, it means the sudden awareness that something unexpected and significant has happened. This surprise response is most likely to be caused by experiences that are unfamiliar or entirely new, difficult to understand, or out of the expected place or sequence.

One of the brain systems that is activated by surprise is the default mode network. As we've noted before, this network is especially engaged by the new, the unfamiliar, and the difficult because it supports just the kinds of complex reasoning functions that are ideal in such settings. So each time a surprising mismatch occurs between our model of the world and our actual experience, the default mode network is engaged to figure out where our model was off and how we might modify it to make it more accurate.

Surprise also activates several other key brain regions. One of these is the hippocampus, which as we've described before plays a key role in learning and memory, two factors that are essential for updating our model. And a third brain center activated by surprise is called the *nucleus accumbens,* which helps provoke our senses of reward, motivation, interest, and salience—the sense that something is especially important and worth attending to.

The process of learning that is activated by failed predictions is called *error-driven learning.* Mistakes usually get a bad rap, and many of us don't like to make them, but the research on error-driven learning actually shows that our brains find the process of learning from surprising mistakes to be inherently rewarding. The resulting surprise stimulates our attention, engagement, motivation, and even sense of reward. This doesn't mean that we will always experience a subjective state of conscious pleasure. Past experiences or outside circumstances can, and often do, render our discoveries of mismatches painful—as when you give a wrong answer in class or discover there was a car coming after stepping into the street. However, as Aristotle said long ago, humans are rational animals, and some of our highest pleasures involve thinking, learning, and understanding. Our brains find the act of testing and refining our mental models to be intrinsically stimulating and rewarding.

Dyslexic Minds Are Optimized to Learn Through Experience . . . and Failure

We believe this error-driven learning process plays an especially important role in the lives of individuals with dyslexia. To understand why, let's recall some of the other things we've already discussed about how dyslexic individuals learn, remember, and build their models of the world.

First, dyslexic individuals tend to form more *lifelike* episodic memories. These memories contain more bits of information that can be used to build better models. Second, dyslexic individuals retain more *separated memories* of particular cases or episodes, rather than *completed patterns* comprised of generalizations. Third, dyslexic individuals form more memories *incidentally* or in a non-task-driven manner. As a result, they store up more information about things whose importance or relevance they do not yet understand, but which may become useful in making predictions in new situations.

This third advantage in incidental learning may in fact be just one reflection of an important difference in the way dyslexic minds are tuned to pay attention to the world around them. Many dyslexic individuals show a shift in attentional balance away from the tight and rigid focus on a carefully selected and rather narrow slice of experience that is considered "normal" toward a more diffuse and less selective awareness of all the things going on around them. Although this more diffuse pattern of attention can lead to challenges in certain situations, there is also evidence that it can lead to advantages in processes such as learning from experience and model building. Let's look more carefully at this topic of focused attention. In chapter 12 we mentioned research from astrophysicist Matt Schneps showing superior visual pattern recognition in dyslexic adults. In these experiments, Matt and his colleagues found that dyslexic individuals showed a more diffuse pattern of visual attention than nondyslexics—that is, they tended to notice more things that lay toward the periphery of their visual fields. They also found that in several different settings, this broader field of attention helped dyslexic adults create a more complete and accurate model of their environment.

Matt's description of the process by which he became interested in studying attentional differences among dyslexics was equally interesting. "I interviewed scientists with dyslexia, and I probed and tried to find out what it was about their dyslexia that they thought might have some bearing on their success as a scientist. One scientist pointed out that he was very good at picking out things that looked out of place, and this really resonated with me. If I see anything that's just kind of 'not right,' I remember it. So to me that was very interesting. That's a hypothesis to start with: Is this something that people with dyslexia have skills for?" Matt's description is an almost perfect restatement of the phenomenon of surprise and error-driven learning.

Attending to Dyslexic Attention

At this point we can begin to see why this process of error-driven learning might be especially beneficial for dyslexic individuals. The amount of learning that takes place through the exercise of this system will increase in direct proportion to the number of predictions an individual makes, and this number of predictions will increase both as the level of detail in an individual's mental models increases and as the scope of their attentional focus broadens. As we've just discussed, dyslexic individuals rate highly on both of these characteristics.

The kind of attentional pattern that will be best suited to this error-driven, experienced-based learning format is very different from that which will be best suited for the kind of rote learning that students are often asked to engage in for their schoolwork. Rote learning is maximized by the ability to narrow one's attentional focus. Students who cannot do this will often be labeled "inattentive," but this is both highly ironic and deeply misleading. These students are actually quite often highly attentive to many aspects of their current experience. It's just that the scope of that attention is broad. Recall the naturalist Jack Laws, who seemed highly distractible in the classroom, where narrow and highly selective focus was required. In contrast, he was able to sit in silence for many hours in the forest, observing the actions of the natural environment as they unfolded around him. The same ability to narrow focus and limit our awareness of most of our environment that benefits rote learning is a significant handicap for error- and experience-based learning. Rather than labeling the attentional tuning of our more broadly attentive students as pathological, we should provide them with learning *experiences* that are more suited to their strengths than simply barraging them with semantic details.

It is also simply not the case that a tight attentional focus is an unlimited good, and a more diffuse focus a handicap. Each has benefits in different settings. For the D-strengths systems we've been speaking of, a somewhat "looser" and broader focus is preferred, and too narrow and tight a focus is a handicap. Research has shown that in most individuals, increasing the activity of the cognitive control systems in the brain's frontal lobes—leading to tighter attentional focus—decreases the activity of their default mode network. As a result, the activities that depend on this network, such as mental simulation, creativity, imagination, and many higher-order problem-solving functions, also decrease. In contrast, in individuals who score highest on tests of creativity, this attention-caused inhibition of default mode network activity does not occur. Instead, these individuals can ramp up both their attention and their default mode network engagement at the same time, giving them an ideal balance of creativity and control.

If you think back to our description of error-driven learning, you can see how important this ability to simultaneously stimulate both attention and default mode activity might be, since the surprise induced by mistaken predictions leads to increases both in attention and in default mode network activity. That's why being somewhat resistant to the usual inhibitory effects of focused attention may be useful for all MIND strengths, as they all rely on default mode network activity. This also aligns with the study we cited in chapter 15, which showed an inverse relationship in Harvard alumni between tight attentional focus and real-world creative achievement.

How Error-Driven Learning Can Correct Our Educational Mistakes

It is simply not possible to overstate the significance of these findings for individuals with dyslexia. Dyslexic minds are quite literally built to delight in learning from experience: from building models of the world, making predictions, and testing those predictions through further interactions and experiences. The implications of this simple truth for the education and working lives of dyslexic individuals are profound, and the often-repeated mistake of treating this bias toward active engagement with the world as a weakness rather than a potential strength is in urgent need of correction.

Dyslexic minds are fundamentally built to gather and use experience in creative and imaginative ways through the application of the MIND strengths. Each individual will vary in the skill with which they employ the different MIND strengths, depending on their abilities and interests. Some will be stronger with spatial or mechanical reasoning, others with creatively linking remote ideas or concepts, and still others with weaving verbal narratives, creating visual arts, solving complex problems, predicting future events, or dealing well with changing situations. The MIND strengths are all built upon a set of common brain systems and perform their functions by remembering and creating detailed mental models of physical and conceptual worlds, both real and imagined, and by using those models for mental simulation, construction, prediction, and other high-level functions.

Dyslexic minds are engaged by the surprising, the new, the difficult, the complex; by tasks that require imagination and problem-solving more than recollection and efficient application of rules or procedures; by deep understanding rather than fluent rote recall; by questions like "Given what you know, what would you predict . . . ?" rather than "What was the name/date/place/definition of . . . ?"

We'll explore the implications of these differences for school and work in the coming chapters.

D-Strengths in Action: Yoky Matsuoka

Throughout this book we've likened the experience of living as an individual with dyslexia to being on a mountain expedition, where different aspects of the dyslexic experience could be likened to different conditions such as slope or temperature. Let's think of visibility. Imagine that your visibility was limited by a fog so thick that it prevented you from even realizing you were on the mountain at all. Then imagine that you spent the first forty-plus years of your life enveloped by that fog, until one day, suddenly, it lifted, and you finally realized where you stood.

For Yoky Matsuoka, pioneering engineer, computer scientist, roboticist, technologist, and entrepreneur, that moment of revelation came in the office of a neuropsychologist, where the third of her four children was being evaluated. "He was in the third grade, and he was very bright, but with his reading he had just completely fallen off the trailer. So we took him to a neuropsych evaluation place, and I was allowed to sit behind him on some of the evaluation. And the very first moment I realized that he and I had the same problem was when they started testing phonemic awareness, or how to manipulate word sounds. And he was having a hard time, and I was having a hard time right behind him. And I thought it was really terrible that I could not manipulate the sounds back and forth. So that's when I realized that I had a problem, and I mentioned this to the guy who was doing the evaluation—I said, 'I couldn't do those problems.' And he said, 'Oh, it's common that some parents find out through these tests that they also have dyslexia.' And I was like, 'Oh, my god . . .'" But then my brain went, 'This really makes sense,' because I had always wondered why as a child growing up in Japan it had been so hard for me to read. I was always so embarrassed to read in front of the class. It seemed like I could do so many other things pretty well, but this was something that I was much worse at than everybody else. I hid as much as I could so I would not be called upon in a classroom, and when they called, I felt like everybody was giggling at me the whole time. And when I was reading or learning English as a language, it was just so much harder for me than for anybody else. I had some problems memorizing my multiplication tables, but beyond that math and physics were so easy for me, like, no problem. But then when it came to digesting letters or for literature, it was so hard. And now it finally all made sense.

"It also made sense of my father. My mom used to yell at my dad for not reading newspapers and say, 'We don't even know anything about the world, because we don't read newspapers!' And it was true, he never read anything. But he was also a very intelligent man who was a successful land developer and urban planner. He came up with new ideas on how to develop cities, and how to rebuild Tokyo after World War II so it would function more like a modern city. He was also a Fulbright scholar, and he went to NYU to study, and then he went back to Japan and was a key developer of Tsukuba Science City, which to this day is a major research center. He was pretty much a typical 'dyslexia idea man'—he could see the big picture.

"I ended up coming to the United States in high school, because I was one of the top-ranked women's junior players in Japan, and I wanted to be a professional tennis player." After spending two years at a prestigious tennis academy, Yoky got a scholarship to the University of California at Berkeley, but a series of injuries ended her competitive tennis career. Fortunately, by that time she had found new passions at Berkeley: math, science, and engineering—particularly the engineering of robots.

After graduating from Berkeley with a degree in electrical engineering and computer science, Yoky enrolled in the graduate program in robotics at the Massachusetts Institute of Technology, where she worked in the laboratory of Rodney Brooks on the famous Cog robot project, whose aim was to create robots that were more humanlike in their abilities to interact with and learn from their environments. Yoky chose to work on the arms and hands, and she threw herself wholeheartedly into the task. "I tend to get really intense and passionate about things, and when I started working on the hand, I began looking around at robots in industry on assembly lines, and they were just horrible. Most state-of-the-art robots used pinching grips to do grasping functions, and they still do. They can be very precise, but they have zero chance of ever becoming a human-level laundry-folding robot. That's how it was in the 1980s when I first started noticing robots, and they're still doing the same thing now in the 2020s. And I thought, as long as we're approaching it in that way, we're never going to get to the point where we are able to build more general-purpose robots—they're trying to solve a fundamentally different problem that has nothing to do with human fingers. And it occurred to me that one of the big problems in the field is that most of the people who were working on it were simply enjoying the math and trying to solve engineering problems, but they weren't really trying to design something that could replicate the hand. And I thought, *What impact is that going to make in the world?* So that's when I said, 'You know what? Forget it. I know that nobody will like what I'm going to do, but you know, I've got to figure out why we have this amazing hand that apparently differentiates us from the monkeys and all the other species so that we could build the tools that built the society and made us so adaptable.' And I thought, *We've got to find out how hands really work, or we'll never be able to build something that works as good.* So that was really my foundation. And then as you said, you know, the more we dug, the more I discovered, and that was so incredible."

Yoky began to study the ways that human arms and hands

actually performed their tasks. She carefully analyzed not only the mechanical components like bones, muscles, ligaments, and tendons, but also the neural systems that directed these mechanical systems. As she did, her work led her deeper and deeper into the study of neuroscience, anatomy, and physiology, and she was fascinated by what she learned. In a way, that was a bit of a shock. "I had always hated biology in high school. It was just meaningless—my teachers didn't connect the information to anything, and I never did well on tests. It was just like geography—just a bunch of lists that I would cram to remember and then forget the next day. But in my studies now I started to understand how the human body could be viewed as a kind of mechanical system, with the brain controlling the physical system, which was living in a physical environment. And I could understand all this using the engineering principles I was already good at and using the same math. Once those connections were there, everything fell into place. And then the more I dug, the more I discovered, and that was so incredible."

After earning her PhD, Yoky spent two years as a postdoctoral student at Harvard, then spent the next decade in faculty positions, first at Carnegie Mellon and then at the University of Washington. Her work flourished, and despite her youth, she became known as one of the pioneers in the new field of neurobotics, which works to bring insights from neuroscience into robotics.

In 2007 Yoky was awarded a MacArthur "genius award" for her remarkable work, but despite her undoubted success, Yoky began to feel out of place in the academic world. "I have always been that funny boundary bridger. It's not that I don't see the boundaries, it's just that I don't understand why people have them. When I got these tenure-track jobs, I was fascinated by neuroscience and fascinated by robotics, and I was publishing in each area. But they were like, 'No, you've got to pick just one, like everyone else.' And I thought that was so silly. They really couldn't figure out what department I belonged in, but I just didn't see it that way. I chased what I needed to know. I mean, electrical engineering and mechanical engineering—the math was the same. One was applied to electrons and the other to physical things, so they felt very interchangeable. And computer science to me was just a tool to be able to do things. And neuroscience and physiology—I could understand them like engineering systems. And to me it just never made sense to try to put boundaries between these. I also realized I was writing papers on things that would be important thirty years from now, but I wanted to make an impact on the world for people right now."

So, in 2009, despite her meteoric rise up the academic ranks, Yoky left academia to became head of innovation at Google, where she was also one of the cofounders of Google X and pioneered the development of the energy-saving Nest thermostat. During the next decade, Yoky held key positions at several other leading tech companies, including Apple, Twitter, and Quanttus. She also held a board position at Hewlett-Packard. Yoky became known as one of the founders of the health tech revolution for her innovative work blending engineering, artificial intelligence, and the biological sciences. As she told us, "I've always felt like whenever I find something interesting to work on, it's going to become a new trend in the world. Like when I combined neuroscience and robotics, and then suddenly there was a field, and it's now one of the most popular engineering majors. And then I combined neuroscience and artificial intelligence and health, and you have health tech, and now it feels like everyone is working on it. I think I've always seen a big enough picture and could just see what the next trend might be—and pouncing on it is really fun."

Most recently, her experience trying to balance life during the stay-at-home days of the COVID epidemic as a mother of four children—two with dyslexia—while simultaneously heading up a new independently led subsidiary of Panasonic and serving as their managing executive director has had Yoky pouncing in a whole new direction. Yoky's new venture, Yohana, matches busy families with an actual human "Yo Assistant," who—backed by sophisticated technology resources—helps with otherwise time-draining tasks like setting appointments, finding and scheduling repair people for the house or the car, scheduling travel, and handling many other tasks. Yoky has also found that the same predictive power that has allowed her to spot trends is very valuable in her management roles as well. "I don't know, again, whether it's dyslexia, but definitely the things that I see are a little bit ahead of the times. So whether it's a personal conflict or who's really going to be able to support a certain part of the organization, I can usually spot it ahead of time."

Yoky also shared a frequent experience that many of our visionary entrepreneurs have related, and which echoes the results of the MIND strengths question on nonlinear thinking. "A lot of people can't understand the things that I can clearly foresee coming in my mind when I describe them. I try to explain what I see, but I realize that the way I'm describing it still may not completely make sense to them. Then months pass, and what I've seen finally happens, and finally they can see it, too, and then we can connect the dots. So sometimes now, even though I wish I could describe it, I just say let's do it, because it can save us a lot of effort. From that point of view, as a CEO, I now get to do a bit of the Steve Jobs thing of like, 'I know you don't believe it yet, but we're doing it this way.' And usually they're fine. But still, it's really interesting how at the higher level some sort of a connection takes place that makes things super obvious to me, but which takes a lot more time to reach logically or even to explain to other people."

This different way of thinking has always made Yoky feel like it was hard for others to really grasp what she was thinking and seeing. "Even when I was a kid in Japan, some of my friends from those days will tell me now, 'You never really fit in here; you were a bit different.' And in college and graduate school, people couldn't really understand what I was talking about a lot of the time. I think that's part of the reason I went into robotics. When I just described things I was thinking, they were like, 'That's an interesting way of thinking about it," but no one seemed to really get it. But when I actually started building things, then they could see the differences in the abilities that the robots had, and they're like, 'Now I totally understand what you meant. And that's a good idea, right?' Because the robots I built were totally different from everybody else's. And that is revolutionizing and moving the needle.

"But for a lot of my life, I felt so different and I never had confidence in myself. I always felt terrible about myself, really. I felt like I was always trying to hide, and always trying to pretend that I was dumb, so that no one knew how hard I was trying. And that really continued all the way until I was in robotics; and then the MIT professors said to me, 'No, you're good. Believe in yourself.' And finally I was like, 'Well, maybe I can. Maybe it's okay.' But it took that long for me to start to believe."

Yoky has decided to use what she has learned about dyslexia and herself to help children who are dyslexic deal with their learning challenges earlier and more effectively, and to build from the beginning the self-confidence that came so late for her. Through her nonprofit, YokyWorks, which she founded using the money that came with her MacArthur prize, Yoky and her family are working to improve early identification and access to intervention for students with dyslexia, to enable them to fulfill their hopes and dreams. It is a beautiful vision from an amazing dyslexic mind who now enjoys the fantastic outlook that her own journey up the mountain has provided.

Part 7

Bringing the Dyslexic Advantage to Life

31

Exploratory Strengths of the Dyslexic Mind

In the previous chapters we've described the MIND strengths, or the common patterns of mental abilities that we have observed in dyslexic individuals. These strengths, rather than challenges with reading, spelling, and other fine-detail skills, are the true core features of dyslexia, and they carry great benefits for those who possess them.

Recently, a remarkable scientist—who introduced herself to us not long after the first edition of this book was released—has published her own theory on the benefits of dyslexia. Using approaches borrowed from a broad array of different disciplines, she has proposed that dyslexic minds create benefits not only for those who have such minds but also for humanity as a whole. In fact, she convincingly argues that dyslexic minds are critical for the ability of the human community as a whole to adapt to a changing and variable world. In the sections that follow, this remarkable scientist—and dyslexic thinker—will share with us the fascinating and important new theory she's developed about the role of dyslexic minds.

Dyslexia and the Theory of Complementary Cognition

Helen Taylor is an archaeologist by training, with a PhD from the University of Cambridge. However, any effort to narrow Helen's full range of interests and expertise to a single professional domain will be entirely futile, because Helen is a perfect example of interconnected reasoning. In addition to a profoundly interdisciplinary mind, Helen also displays a deep intellectual tenacity and passion to pursue the questions that interest her.

In recent years, the questions that have most sparked Helen's interest relate to dyslexic minds. Helen has developed a detailed theory that casts new light on the value of dyslexic minds both for those who have them and for humanity as a whole. Helen's theory of *complementary cognition* is rooted in the understanding that humans adapt primarily through culture, by exploring, refining, and consolidating behaviors and inventions that contribute to our survival.

Helen's theory states that humans are cognitively specialized in different but complementary learning strategies that underlie this process of cultural adaptation. These strategies lie along a continuum from global search, or *exploration* of unknown information, to local search, or *exploitation* of existing knowledge. To adapt, humans must identify problems in current knowledge and update their behavior and inventions. Helen's theory proposes that collaboration among individuals who fall across this continuum of learning strategies (from exploration to exploitation) results in the creation of higher-quality solutions to the challenges of adaptation—solutions not otherwise obtainable from an individual or from a group of similar individuals. This superior effectiveness of collaboration makes us interdependent; people with different abilities must work effectively together to adapt and solve problems successfully.

In a recent research article, Helen outlined evidence showing that people with dyslexia specialize in exploration. This enhanced ability comes as a trade-off to exploitation-related skills, such as those needed in reading and writing (e.g., fluent application of rules for decoding, spelling, syntax, and mechanics). Therefore, within the theoretical framework of complementary cognition, dyslexia can be viewed not as a disorder but rather as a mismatch between a learning strategy and the ability to read and write.

To help explain the theory in more depth, we asked Helen to describe to us how she developed her ideas. Helen's interest in dyslexia had its origins in her own experiences as a dyslexic person. Though she was not formally diagnosed until college, Helen struggled throughout her education, with difficulties in reading, writing, orienting letters correctly, listening to lectures, organizing materials, managing her time, memorizing by rote, and in general learning anything in the way it was taught or demonstrating her knowledge in the expected forms. It was only by using every available moment (including lunch breaks) to do her schoolwork and gradually discovering how to teach herself to understand what she was learning (often through the use of drawings) that she was able to get by.

However, when she eventually reached the university level, the demands from her coursework exceeded her strategies for coping, and she finally received psychoeducational testing. The diagnosis of dyslexia that Helen received provided her with an immediate explanation for her struggles; but over the months that followed it led slowly to a recognition that this explanation was both incomplete and misleading.

"When I first got diagnosed with dyslexia, I was struggling so much that I thought I had something seriously wrong with me. It wasn't until a little while later when I was studying the invention of writing that I began to question the idea of dyslexia as a disability. I realized that writing had always been my biggest problem—I just can't get my ideas into words or into an ordered structure, and it drives me insane; but I also realized that writing was primarily a problem because I couldn't get things that in my mind were all interconnected in three dimensions down into a linear sequence on paper. And I thought, *If that's my problem, then it's a strange sort of problem,* because I also realized that my thoughts were often quite original and that I can always figure things out, and the reason I could was because I could make all these connections. So I thought, *How can I have a disability if the thing that's making writing so difficult is also the thing that's helping me figure things out?*

"Also, I began to wonder how we could consider a difficulty using a technology to be a disability—because that's what writing is: it's a very recent human invention. Written language is really the only cultural invention for which we assume that difficulty with use constitutes some kind of deficit. If someone isn't good at, say, accounting or computer programming, we don't assume they have a neurobiological disorder. But we do that with reading and writing.

"So that's when I started questioning the whole idea that dyslexia should be seen as a disability, and I began to look for a better way to think about it. At that time—around the year 2000—the only alternative view I could find was an article on visual communication in art students. Instead, I started to question all my friends and I asked them what was going on inside their heads. I'd give them an activity and I'd ask, 'What is your brain doing? What do you see?' Or, 'If you come across this problem, how do you approach it?' I just interrogated them on how their brains worked for years, and I started forming my own ideas on what some of the different strengths were, including the strengths that dyslexic people seemed to have, and made lists of them. And about that time your book [*The Dyslexic Advantage*] also came out, and that was also very exciting because it aligned so well with the ideas I'd been forming. For example, your *interconnected reasoning* was precisely what I had been calling *networked thinking*.

"In the process of all this questioning, one of the things that I discovered was that my best friend was in many ways the opposite of me. She had a much more schematic view of things, so when I asked her about what she thought about when I said *brain*, she said she thought about a simple grayscale drawing of a brain, whereas my mental image was much more like a real brain, with color and weight and texture and all those other features. She is very detail-oriented, and this was reflected in her talents for writing, art, and music: she paints very realistic pictures and composes and plays fast guitar solos that require hours and hours of practice. She also didn't have any problems with reading. But then she'd struggle with things I found easy. For example, she'd get lost very easily because she had no mental map of where she'd been, whereas once I've walked somewhere, I just know my way around and quickly develop a map of the landscape so I can simulate a fly-through of the area and decide what would be the best route. So there were all these things one of us could do and the other couldn't, and vice versa, so that we made a very good team, but we were also very different. And I realized in fact that we were *complementary*, and it really struck me that perhaps this kind of complementarity among people wasn't simply an accident, but that it might point to something very important.

"I mentioned this idea of specialization to my friend Pablo, who is an economist, and he gave me a paper that discussed the benefits of the division and specialization of labor and the importance of effective cooperation. That became a very important part of my thinking. But this idea of specialized thinkers also caused me a problem because I thought, *If we're specialized, then what are we specialized* in?

"So I tried to find a pattern among all the different strengths I had identified and read about, and I spent ages thinking about that. But my mind also kept going back to that economics paper, because it discussed not just one but two key factors in improving productivity and efficiency. The first, which was the one I had thought primarily about, was specialization and collaboration. But the second was increasing general knowledge—and that's what eventually gave me the clue. I realized that the dyslexic strengths on the lists I had made all dealt in one way or another with seeking knowledge about things that are *unknown*. So I thought, *Maybe that's what dyslexic strengths are focused on—seeking and dealing with the unknown, as opposed to focusing on capitalizing on the known.*

"I began to wonder: How do I verify whether people with dyslexia are specialized in exploring new opportunities? Then one day when I was walking, I realized that this balance between using existing and creating new information is what we see in adaptive evolution itself, with a small number of mutations to provide new opportunities within the larger background of stability. So I continued to look for more examples of known/unknown trade-offs in systems that adapt until eventually I came across complex systems theory. Then I read a book by Scott Page where he said, 'The trade-off between exploration and exploitation arises often in complex systems,' and I thought, *That's it*—that's *what I'm talking about!*"

In complex systems theory, the idea of *exploration* refers to activities that are oriented toward searching for new opportunities, resources, or rewards, while *exploitation* refers to sticking with opportunities and resources that are already known and have so far proved rewarding. Maintaining an optimal balance between exploration and exploitation at the whole system level is critical for successful adaptation and survival. As Helen has written: "Exploring endlessly without exploiting what has been found can be inefficient, whereas focusing too much on exploitation may be suboptimal or result in failure to adapt to change. This trade-off arises in many seemingly unrelated areas of endeavor, from evolution to the economy to artificial intelligence."

Helen went on: "As I read through these papers, I learned that in complex self-adapting systems, this exploration-exploitation complementarity is seen as particularly useful in enabling systems to adapt to settings where the environment was especially variable." One well-studied example is the business environment, where research has examined the beneficial effects of mixing exploratory traits like innovation and experimentation with exploitative ones like increasing efficiency and focus for success.

"Applying these ideas to my earlier research, I realized that one thing that complex human systems definitely need in order to adapt to variable and uncertain environments is to be continually exploring and updating their information, to ensure that they're maintaining the optimum level of exploitation. And to do that, it is clear that at least some of their members will need to be focused on what is currently unknown." Helen's first goal was to ensure her ideas could avoid one obvious disconfirmation. "I created a testable hypothesis: If exploratory cognition really is an adaptive response to variability, then we should expect to find that the environment in which humans evolved was highly variable. To test that hypothesis, I started looking back at the paleoenvironmental evidence, and what I found was that the most recent six million years—and especially the last 400,000, during which humans evolved—have seen the most dramatic climate variability of any period since the extinction of the dinosaurs. So that aligned perfectly with my prediction.

"Now I really felt that I was onto something, and my next step was to determine at a cognitive level whether there was any evidence supporting the idea that dyslexic brains were, in fact, geared toward exploration. Since the brain is also a complex adaptive system, I looked into the literature to see whether anyone had studied cognition from that perspective, and that's how I found the field of cognitive search." Cognitive search investigates how our minds (or more recently, artificially intelligent machines) look for resources relevant to our current interests or needs either in our memory stores or in the outside world. "I studied that work as well as all the research on dyslexia in different areas of cognition to see whether it all aligned with this exploratory pattern, and I found that the evidence very much supported the idea that dyslexia is an exploratory specialization, which is oriented toward seeking the unknown. For example:

- The lower working memory capacity most individuals with dyslexia show tends to be associated with a preference for global or large-scale search strategies rather than local or fine-detail exploration.

- Lower working memory is also associated with insight-based problem-solving, which is closely related to exploratory search.

- Enhanced divergent creativity, like that found in dyslexic adults, is beneficial for exploratory behavior.

- By promoting the retention of conscious access to information [i.e., conscious compensation], the automatization deficit found in many people with dyslexia [see chapter 5] may make it easier to modify and integrate information, thereby facilitating innovation.

- The broader and less focused fields of sensory attention found in dyslexic individuals in areas like vision and hearing favor global rather than local search.

- The use of episodic memory systems for mental simulation is inherently exploratory in nature.

- The broader minicolumn spacing found in dyslexic individuals favors global over local processing, and exploratory behavior.

"So my conclusion based on all this data was that the exploratory model of dyslexic cognition, with corresponding trade-offs in exploitation or local processing, holds up very well.

"I think this pattern of complementary cognition is so important, because the trade-off between exploration and exploitation that has resulted in different but complementary cognitive specializations among humans hasn't just been critical to our ability as a species to adapt to change in the past—it's equally essential for our present and future. This complementarity is simply not something we can ignore or neglect, because once a system's components have specialized like this, they become interdependent, so all ways of thinking become essential both to the individual and to the health and success of the whole. As a result of our exploitation-biased education system, we might be failing to recognize and understand the value of exploratory cognition, which will not only harm individuals with dyslexia and limit their opportunities, but will also weaken our collective ability as a species to thrive and adapt to change.

"In the long run it will be counterproductive if we continue to think of cognition exclusively in individual terms and insist that everyone become optimized for more exploitation-related skills. Instead, we need to think about cognitive search at the group level as well, and recognize that some people are simply tuned more toward exploring the unknown than exploiting existing knowledge—and that is a good thing, not a disability. It's also what has made the human species so adaptable. . . . Ultimately, humans adapt to change primarily through culture, and if our cultures are to remain adaptive, they must support the growth and development of individuals who think in different but complementary ways. Ultimately we humans are *interdependent*, and that means that whatever learning strategy we excel in, we must all collaborate effectively to collectively adapt and solve our problems."

Exploratory Strengths in Action: Bob Ballard

Dr. Robert Ballard is universally regarded as one of the great explorers of our age. Bob is best known to the general public as the discoverer of the resting place of the *Titanic,* as well as the finder of many other famous shipwrecks, including the German battleship *Bismarck,* John F. Kennedy's PT-109, and the *Lusitania,* a passenger ship whose sinking in 1915 by a German U-boat played an important role in drawing the United States into involvement in World War I. Yet these famous discoveries are only a minor part of Bob's contributions to our understanding of the deep seas. Early in his career, Bob provided definitive evidence in support of the theory of plate tectonics, revolutionizing geology. He discovered the first deep-sea hydrothermal vents, and with them stunning new life-forms that have opened up entirely new fields in biology and biochemistry. He has combined oceanographic and archaeological techniques to find and explore shipwrecks in the Mediterranean and Black Seas from as far back as the Canaanite era (c. 2000 BC), helping to rewrite the history of ancient navigation and commerce. And he has used his military background and knowledge of the topography of the ocean floor to make major contributions to undersea defense strategy and submarine warfare.

When looking over Bob's list of accomplishments—or better yet, reading (or listening to) his fascinating autobiography, *Into the Deep*—we find it almost impossible to believe that this list describes the work of a single person. It is even harder for many people to believe that a person so remarkable has written about himself: "I've always felt insecure, deep down, about my brains and ability. Even with all the success I've had I've never been able to shake the fear that I was like a boxer trying to fight opponents in a higher weight class."

Bob attributes this lifelong insecurity to the problems with reading and writing he experienced as a student, and to the difficulty he felt in "measuring up" to his academically talented older brother. He was told he was "hyperkinetic," but it felt to him as if school were a sort of straitjacket, and he struggled from the beginning to keep up with the required reading.

When his family moved from Kansas to California and Bob entered fifth grade, it was clear he was far behind. His mother worked with him to build his organizational skills and improve his focus, but he still received a D for reading on his report card. Bob's parents arranged for him to have daily tutoring, and while that began to help, what really caused him to buckle down and improve his school performance was an event that gave him something that tutoring couldn't: motivation. At the age of twelve, Bob attended a showing of Disney's film adaptation of Jules Verne's *20,000 Leagues Under the Sea*. That experience changed his life. He realized in a way he never had before that there was a vast, three-dimensional world beneath the ocean's surface filled with unknown landscapes and creatures, just waiting to be explored. He suddenly knew what he wanted to do with his life: He wanted to be Captain Nemo! Bob now had a dream, and with that dream came the determination to do whatever was necessary to achieve it. He realized he would need good grades to pursue his goals, so he began to work as he never had before, and his grades began to rise.

In high school, Bob volunteered during his summers to work with oceanographers from the Scripps Institution of Oceanography. Those experiences helped him confirm that oceanography was his path. He enrolled at the University of California at Santa Barbara, beginning an ambitious five-year program in physical sciences that appealed to his broadly interdisciplinary sensibilities. He hoped the degree would make him an attractive candidate for later admission to the PhD program at Scripps. However, his continued problems with reading and test-taking, which held his SAT scores below 1000, also caused problems in some of his classes. In classes like calculus and organic chemistry, he had trouble picturing in his mind the things his textbooks described in words, and he received low marks in these classes. By the time he finished his degree he had pulled his grades up to a 3.0 average, but this was not high enough for Scripps. Bob's application was rejected.

After a crushing bout of despair at this failure to achieve a long-cherished dream, Bob picked himself up and began looking for plan B. He enrolled in a master's degree program at the

University of Hawaii. Later in life, Bob looked back on this experience as one of the most valuable in his life. He realized that failure could be a learning lesson. "Since then, I've grown to love failure. It's not something you should try to avoid, but rather embrace and learn from—and then beat. You don't go around it—you go through it. You get knocked down, and you lie there, and you go, Wow, that was a hell of a shot. And then you dive back in. I've failed lots of times, but I don't quit, so I always win. . . . I've learned to embrace failure as the greatest teacher."

After two years Bob returned to California and enrolled in a doctoral program in marine geology at the University of Southern California. It was 1967 and the Vietnam War was at its height, so soon Bob, who had been an Army reservist in college, was called up to active military service. Given his training in oceanography, Bob received a transfer to the Navy and was assigned as a liaison between the Office of Naval Research and the Woods Hole Oceanographic Institution in Massachusetts. At Woods Hole, Bob began work on *Alvin*, a recently developed manned submersible deep-sea research vehicle (submarine) that was beginning to open up new and previously undreamed-of opportunities for undersea research and exploration. He was Captain Nemo at last!

Over the next few years, Bob became a tireless advocate for using *Alvin* in a wide variety of scientific and defense-related applications, and he was allowed to pursue a series of explorations that led to the remarkable series of findings mentioned earlier. In these years, the boundless energy, curiosity, and enthusiasm that made it hard for him to focus inside a classroom or in front of a textbook made him an ideal self-starter and problem-solver. "Once my career got going, I was always in motion, barreling from one project to another. By then, I believed that with hyperfocus and energy, I could overcome any challenge."

It was not only Bob's energy and focus that led to his success. His highly interdisciplinary mind, deeply collaborative spirit, and insatiable desire to be always exploring the unknown—whether by visiting uncharted places or pushing for the development of new technologies—also played key roles. As he told us, "The most exciting place in science is right at the cutting edge—at the place where the cleared field meets the dark forest—and especially when two edges touch. That's where the lightning is. I think a lot of people are terrified of freedom—of the idea that no one's taking care of you. Sometimes I'll ask my students to define a vacuum, and one will say, 'It's a free space where you can expand in any direction,' and another will say, 'It's the absence of air, so you'll suffocate.' Not everyone likes that freedom! But I think we dyslexics are used to living alone in that free space, so we're not afraid of it. We're comfortable out exploring—and in fact we love those dark and lonely places—because that's where the excitement is, out where most people are afraid.

"I am an explorer and a wanderer. When you're dyslexic, you're an outsider, and so you learn to wander outside the system, . . . but you can get into trouble because you tend to wander into other people's territory. Every group has these people that I call 'the guardians of the difference,' because they think their job is to divide society up into bins. And the key is, can you get past the gatekeepers to get to the good people? I remember when I was at Stanford on sabbatical for a couple of years. Allan Cox was the dean of the College of Earth Sciences. They had about nine different departments, and Allan told me, 'I've tracked the students that go through this system, and the ones that do well in life are the ones that go to seminars in departments other than their own to get new ideas.' It's sort of like Reese's Peanut Butter Cup, where the peanut butter runs into the chocolate. My greatest successes have come through mixing different fields."

In addition to his highly interconnected thinking style, Bob also benefited from strengths in spatial reasoning, mental simulation, and prediction. These have also played a critical role in leading to his success in the "needle in a haystack" searches for undersea shipwrecks where so many others have failed. For example, the insight that led to his finding of the *Titanic* resulted from a process of mental simulation in which Bob began with eyewitness accounts that described the ship as breaking into two large pieces before sinking. Bob imagined in his mind what would happen to the wreck's fragments during the more than two-mile descent to the ocean bottom. He saw like a film in his mind that the heaviest pieces would go essentially straight down, but that the lighter bits—the ship's furnishings, baggage, and other pieces—would be pushed by the ocean current and would settle out like a lengthy comet's tail on the ocean floor. As these images played out in his mind, Bob suddenly realized that instead of searching for the larger fragments of the ship, whose combined lengths would total at most 883 feet, he should be searching for a debris trail, which would be a mile or more in length. This would vastly improve his chances of finding the *Titanic*—and that was exactly what happened.

Even after he'd begun to assemble his long list of accomplishments, Bob's nonlinear thinking style, based on mental simulation and prediction, often led him to think in ways that others could not easily follow. "I love how people always say, 'This idea you've got is crazy, though your last one was pretty good.' And I say, 'But don't you remember that you called that one crazy back then, too?'"

Despite his many successes, one mystery continued to puzzle

Bob: the mystery of his own mind. He knew he differed greatly from most others around him in the more visual way he approached his scientific research and his shipwreck searches, but he never understood the source of this difference. Nor did he understand how a mind that struggled so hard to read and to channel its energy in the classroom could be so ideally suited to succeeding in so many complex real-world endeavors. The answer came, quite literally, from out of the blue: "One day in March 2015, I was driving home from my office, and I heard a segment on the radio about a book called *The Dyslexic Advantage*. What I heard felt familiar. Could I have been dyslexic without even realizing it? I ordered the book that night, and when I began reading it, I couldn't put it down. Tears were streaming down my face. Here I was, 72 years old, and this book, finally, was explaining me to me. Even now I think of it as my first autobiography. That's how much it mirrored who I am."

As Bob told us recently, "Through your book and through my self-analysis, I've really embraced being dyslexic. 'Dyslexic advantage'—just seeing those two words close to one another made a statement in its own right, and that's what's so wonderful. I always knew I was different, but it was through hearing your interview on NPR and then reading your book that really showed me I wasn't alone. There's a lot of us out there—what an untapped resource! And it's made me even more dangerous, because now I really know what I'm good at and what I'm not good at, and I'm homing in on what I'm good at and having a ball dreaming up stuff. And it's also helped me realize why I've always loved working in teams. My wife, Barbara, is a producer, and she handles all the production details for our National Geographic specials and other projects, and negotiates contracts with a skill I could never muster, and together we've become a sort of dynamic duo of exploration. Until recently I've also had five full-time administrative assistants who help me with the day-to-day details in each of my major project areas. I have found that I can't function without nondyslexics. I need nondyslexics to implement, and without that I cannot operate. We all work best together.

"I wish everyone recognized that dyslexia has its advantages, and the best education is the one that brings those talents out in a child. For dyslexics, the printing press was our demise as far as education was concerned, because before that learning was primarily through oral communication and doing and demonstration and visualization. We need to be mobile. We need oral testing and other ways of evaluating us and letting us survive the educational system. I just survived by trickery—memorizing and time management and all that nonsense—but we need something better: We need an educational system that's sympathetic and understanding about dyslexia. And it's not even the Bob Ballards who are worst off. It's the people of color who are dyslexic that are being really trashed with high dropout and incarceration rates because they end up as entrepreneurs in the wrong kind of business. It costs more to send a kid to prison than to Harvard, and they only go to Harvard for four years. Some of these kids are incarcerated their whole life, and the economic loss to our society of incarceration of dyslexics is staggering. That's where the flashlight needs to be turned on—big and bright—particularly given our greater awareness now in our society of marginalized people.

"I'm always on the lookout now for other dyslexics, especially children, and I encourage their parents and teachers to nurture the gifts they have. When I give talks to kids, I'll say at the beginning that I'm dyslexic and ask if any of them are. No one raises a hand. But by the time I finish my presentation, I ask them again, 'How many of you are dyslexic?' And *now* they raise their hands! They have learned that someone just like them has followed his passion and gone on to have a successful life. That is what I want them to remember. Not that I found *Titanic*, but that I set goals and kept working to achieve them—and that my dyslexia actually helped me get to where I am today.

"We're all on a journey. The explorer in me loves the challenge, the chase, the search, the raw energy, and the rush—the thrill of finding something you never set out to find as well as the sweet victory of discovering what you were looking for all along. Whenever anyone asks me what was my favorite discovery, I always answer with a phrase that I've used again and again: 'It's the one I'm about to make.' I still feel that way, and even though I'm turning eighty next month, I'm headed off to sea again to do what I do best: explore."

32

Educating Dyslexic MINDs

Given what we've learned about dyslexic minds—how they develop, where they show strengths, where they struggle—and the important role that they play in human society, what are the implications for the education, employment, and emotional well-being of individuals with dyslexia?

In the final chapters of this book, we'll examine those questions, beginning in this chapter with a focus on education. Typically, when books on dyslexia turn to the topic of education, they focus on the nuts-and-bolts details of reading and spelling instruction. That's important and useful information. In fact, it's the kind of information we provided in the first edition of this book. This time, however, we'd like to offer a bigger-picture perspective on dyslexic education. In particular, we'll look at the implications of the dyslexia-associated cognitive patterns we've discussed so far—both strengths and weaknesses—for learning and education more generally. For those who are still interested in the type of more specific and detail-oriented information we provided in the first edition on these and other challenges to education, such information can be found at our nonprofit website, dyslexicadvantage.org. For everyone else, we'll invite you to pull back your focus and think with us about what an education that's designed for dyslexic minds might look like.

Educating Dyslexic MINDs: It's Time for a New Approach

In recent years there has been a remarkable growth in efforts aimed at improving the educational outcomes and experiences of dyslexic students. Here in the United States, laws have been passed in many states mandating early screening to identify students at risk for dyslexic reading and spelling challenges, and also better training for teachers in dyslexia remediation.

However, as we have said many times, dyslexic minds don't just differ in the ways they learn to read and spell. The differences that distinguish dyslexic from nondyslexic brains permeate every aspect of learning, attending, reasoning, and expressing understanding and knowledge. Therefore, as commendable as these efforts are, any attempt to address the educational needs of dyslexic students that fails to recognize the radical and comprehensive nature of these differences—and to provide alternative instructional approaches designed to nurture minds that work in these different ways—will fail to create an educational environment that can promote the growth of dyslexic minds.

We are not in any way slighting the importance of early identification and reading remediation. As research has shown, reading remediation is in many important respects easier and more effective when begun in students before age nine, and improving the reading ability of dyslexic students is unquestionably a good thing. However, we will not truly meet the educational needs of our dyslexic students if we teach them to read just so we can give them lists of facts to memorize, then assess them with multiple-choice tests. A "reading patch" won't fix the existing systems. We need to construct a new approach.

We should begin by reflecting carefully on how dyslexic students differ in the ways that they learn, think, and develop, then create learning opportunities and environments that can better nurture the positive differences while supporting and remediating the challenges. The stakes in providing an appropriate education for the lives and futures of these dyslexic students, and for society as a whole, are profound, and the magnitude of our response should be scaled accordingly.

In the rest of this chapter, we'll discuss what a truly dyslexia-appropriate education should look like.

Big-Picture Principles for Educating Dyslexic MINDs

Let's approach the task of designing a dyslexia-appropriate education like a dyslexic mind would: by starting with the big picture and identifying the key principles that should guide our choices. In the section that follows we'll describe three such principles: experience, understanding, and prediction.

Experience

We base our selection of the first principle, *experience*, on two observations. The first is that dyslexic minds are strongly biased toward episodic or personal forms of memory. Dyslexic students are more likely to remember things they've experienced—or things they've imagined as experiences—than they are to remember other types of information.

The second observation is that dyslexic MIND strengths are powered by systems that use the components of these episodic memories as building blocks for creating models of the physical and conceptual worlds and for constructing all sorts of mental scenes and simulations. The more plentifully a dyslexic mind is stocked with such fragments of experience, the more productively it will be able to use and benefit from these constructive systems.

Firsthand experiences are an excellent way of stocking the mind's storehouses with episodic memories because they provide so many opportunities for incidental as well as intentional learning. The most memorable experiences will be those with an element of surprise, novelty, or personal relevance. Things that are commonly observed in the world but whose causes or purposes may have remained hidden can be great starting points for exploration (e.g., Why is the sky blue? Where does the wind come from? How do we get electricity? Who pays for schools? Who makes our clothes?).

For some students with spatial, mechanical, or artistic interests, the most interesting experiences may involve creating, building, or taking apart. For other students with a more verbal or interpersonal focus, interesting experiences might involve holding a debate, mediating a dispute, engaging in conversations, or negotiating. Creating a new story, providing an alternate outcome for an existing story or historical event, or engaging in a symbolic or historical reenactment can also aid learning. Many students might benefit from field trips or site visits, preferably after receiving enough background information to prepare them to ask interesting questions. Or these interesting experiences might simply involve unstructured time for exploration, mind-wandering, experimentation, and insight.

Episodic-type memories can also be formed secondhand by hearing stories or cases that present information in the form of an experience that can be imagined or mentally simulated by a student with a rich episodic scene construction system. In this way students can imaginatively share the experiences of others, making those experiences in a sense their own. Students can even learn to transform relatively hard-to-remember abstract, impersonal, or semantic-type information into more memorable episodic-type forms by imaginatively creating experience-like memories that creatively weave the semantic information into their narrative. Such approaches are called *mnemonic* (or memory) strategies, and they are the source of the record-setting memory power that experts like Kevin Horsley use and recommend. Exposing dyslexic students to different mnemonic strategies (visualizing journeys with facts tied to particular places; creating clever drawings; inventing stories; using rhymes, raps, or acrostics; etc.) and instructing students in the forms that seem most effective for them should be one of the core activities of any dyslexic education. By boosting memory efficiency through the use of well-functioning episodic systems, students who fail disastrously at rote memorization can become not only adequate but truly excellent rememberers of all sorts of information.

Bottom line: One of the main goals of dyslexic education should be to equip dyslexic students with elements of experience that they can use and reconfigure in creative, imaginative, and innovative ways. When considering the kinds of educational experiences to which dyslexic students should be exposed, always remember that the long-term goal is to equip them to be excellent mental simulators, problem-solvers, predictors, and explorers of the unknown, rather than repositories of known facts.

Understanding

The second principle, *understanding*, reflects the contextual, big-picture, and interconnected nature of knowledge that characterizes dyslexic minds. In our I-strengths survey, over 90 percent of dyslexic adults agreed that to learn anything new, they had to connect it with something else they already knew or to place it in a big-picture context. Many of the dyslexic individuals profiled in this book have also described how dependent their learning was on understanding underlying principles and context, because rote approaches simply did not work. This kind of true understanding rather than simple fact assimilation should be an aim of dyslexic education.

Viewed properly, an inability to retain information not connected to some larger network or structure of understanding has a great potential benefit. While it can be inconvenient for tasks where rote memory comes in handy—like trying to remember phone numbers or the longest rivers in India or the rules for dividing and multiplying fractions—the flip-side benefit is that you are less likely to fool yourself into believing that you understand something simply because you have memorized a definition, date, or procedure. Being unable to remember something until you truly understand and "own" it—see its context, relevance, utility, and applications, and grasp how it connects with other things—is in fact a gift, and it's what we should be aiming for in dyslexic education.

Most often this understanding will come through seeing the deeper principles underlying a fact or concept as displayed in action—that is, through witnessing, imagining, and experiencing them. Examples and analogies from multiple cases and disciplines can build the connections that create true understanding. Remember that when you are trying to teach a dyslexic student something new, you should never start from ground zero or a clean slate. Always begin with something that is familiar and already known, then build bridges to the new knowledge. Building this sense of interconnected knowledge will help dyslexic learners feel oriented even as they head out to explore the unknown. Similarly, starting lessons or reading assignments with a short preview or road map laying out the big picture and the final destination will also improve understanding and retention.

Prediction

Our third principle, *prediction*, reflects what we have described as the primary learning style of the dyslexic mind: error-driven learning. As we've explained, dyslexic minds excel in building models of the physical and conceptual worlds using hippocampal grid-type networks and the episodic simulation system. Moment by moment, these models are used to make predictions about what will happen next. When these predictions are confirmed, little notice is taken and further predictions are made. When a prediction fails, the mind experiences surprise, attention is drawn to the error, and efforts are made to understand the error, then correct the model. As we've discussed, the brain experiences such surprises as inherently interesting and even rewarding, and this heightens the effectiveness of learning.

When instructing dyslexic students, we should create such error-driven learning experiences as often as possible and sprinkle them throughout the curriculum. In addition to taking advantage of what exploratory dyslexic minds are optimized to do, repeated experience with error-driven learning approaches can destigmatize the making of mistakes, recasting them instead as excellent opportunities to learn, understand, refine ideas, and do better in the future. This kind of attitude toward learning from mistakes has often been cited by successful dyslexic adults as one of the keys to their flourishing despite setbacks.

Giving prediction a more prominent place in education would not require a radical restructuring of the curriculum, just a change in how information is presented. Instead of passing out bits of semantic information like little pills to be swallowed, students should be shown cases or examples, asked to infer principles, then asked to predict how these principles might operate in other situations or contexts. They might be asked to shift their perspective or point of view, imagine an alternative scenario or a hypothetical situation, or speculate on the possible uses of a bit of information. The point is to draw students into an exploratory process of learning, in which they are making predictions, testing them, and refining their mental models, rather than just receiving information passively.

Error-driven learning is so essential to how dyslexic minds learn about the world that we might think about it as a *dyslexic learning loop.* To make the steps of this loop more memorable we can use the acronym IDEA to remember them:

1. *Imagine*: Using the models we've created with our mental simulation and problem-solving systems out of bits of past experience, we *imagine* possible outcomes or scenarios for what we expect to encounter either immediately in experience (small-p predictions) or at some point (big-P predictions).

2. *Decide*: Using those same systems, we *decide* on the most likely outcome and take that as our prediction.

3. *Explore*: We then move forward to *explore* the physical or conceptual worlds where we put our predictions to the test.

4. *Assess*: We *assess* the results of our predictions. If our prediction is confirmed, we move back to step 1 and continue the cycle with a new prediction. If it is not correct, we seek for the source of the error, refine our models, then go back to step 1 to continue the cycle.

Reading and Learning

It is hardly possible to have a chapter on dyslexic education without at least mentioning reading; but since our focus in this chapter is directed more on big-picture principles than on details of instruction, our consideration of reading will be big picture as well. In particular, what we'd like to consider is not how to make dyslexic students into better readers—which is a very important topic, but one well covered in many other places—but what we should do for those dyslexic students whose reading skills are not yet strong enough to allow them to perform up to the level of their full cognitive potential.

When we consider the place of reading in schools, it is common to say that up through age nine or so, students are in the "learn to read" phase. In this phase, reading is treated as its own subject. The focus of instruction is to master reading as a skill, rather than to use it as a tool to access information. Once students reach age nine or so, everything changes. In the new phase that follows, students are expected to read to learn. Reading skills are no longer taught but are assumed to have been previously developed. Throughout the curriculum, students are now expected to use competent reading skills to access all sorts of information, and those whose reading skills are not up to the task will increasingly face hardships. Dyslexic students with the poorest reading skills typically face a cycle of increasing failure. Those with partially remediated challenges may avoid failing outright, but they often fail to perform up to the level of their general mental ability. This can result in chronic underperformance, with a resulting loss of interest and motivation. Very bright dyslexic students may get by in the general classroom but fail to keep up in gifted or accelerated programs due to an inability to keep up with reading demands; or they may get by on a day-to-day basis, only to have their reading problems derail them on tests that are taken under time-limited conditions.

The cycle of problems that occurs when students are not ready for this transition is well documented. The typical response of educators and advocates has been to stress the importance of improved reading instruction prior to this transition at age nine. Such an emphasis is unquestionably important, but it is not a complete solution. No matter what we do to improve reading instruction, there will still be students who do not read up to grade level (and many more who do not read up to the level of their general cognitive ability) by age nine, and we are currently failing to provide good educational alternatives for these students.

It is also a serious mistake to conclude, as many somehow have, that reading progress can be achieved only up to age nine. This misunderstanding creates a "countdown to doomsday" mentality for reading instruction that is as misleading as it is harmful. The following egregious example, which appeared in a newspaper editorial, is one of many we could quote: "Children who do not learn to read and write by the end of third grade are likely doomed to school and life failure that can often lead to drug use and criminal behavior." It is true that poorer outcomes of various sorts are statistically more likely for those who have not learned to read at grade level by age nine. However, these poor outcomes are better interpreted as an indictment of our current failure to support late-reading dyslexic learners than as proof of an inherent connection between poor decoding at age ten and an aptitude for social deviance. It is simply not true that students who have not learned to read by age nine cannot be successfully educated, even if their reading remains problematic. In fact, nearly all of the successful dyslexic persons profiled in this book did not read at grade level by age nine, and many could not read at all until well after that. Several of the most brilliant people profiled in this book have told us privately that they have still never read a single book cover to cover. Others like Vince Flynn and Kevin Horsley were well into their twenties before they tutored themselves into becoming skilled and voracious readers— and in Kevin's case, the author of a book on speed-reading. And students like Jack Laws and Blake Charlton cited their dependence on text-to-speech readers and recorded books to keep up with reading. Simply providing better access to such assistive technologies for students who do not yet read as well as they think can prevent underperformance due to an inability to access information from texts. As important as it is to improve reading ability, there is simply no good reason to deny access to assistive technologies for students whose performance is being hindered by their lack of skilled reading.

A Truly *Equal* Education

Ultimately, because the cognitive differences that characterize dyslexic minds affect essentially every aspect of learning and development, dyslexic students would do best in educational programs that are designed with their unique developmental and cognitive differences in mind. But there is a lot that can be done to incorporate more dyslexia-friendly approaches into all educational settings without major curricular changes. First, we should provide universal literacy instruction that also meets the needs of dyslexic students. As Professor Maryanne Wolf, director of the UCLA Center for Dyslexia, Diverse Learners, and Social Justice, told us, "For a long time there has been a polarization in reading instruction. On the one hand, we have those who believe in whole language and balanced literacy approaches, who believe they are supporting the imagination and inductive reasoning of children. On the other, we have proponents of structured literacy, who believe that many children will never have a chance to become skilled readers without systematic instruction on the foundational skills that form the lowest rungs of the reading ladder.* In fact, good reading instruction must be comprehensive in both. In our work, we are trying to help teachers realize that if you understand how to teach the different strengths and weaknesses within dyslexia, you will have better tools for teaching all children. If you understand dyslexia, as a teacher, you will have a better way of teaching all children."

Second, we should emphasize experience, understanding, and prediction throughout the curriculum. With a little forethought and planning, these principles can be made a part of most lessons. As with the reading instruction described above—and as the MIND strengths surveys show—these changes would not be valuable just to dyslexic students. The primary difference is simply that for the dyslexic students, it is often a matter of success or failure, rather than a mere learning preference.

Magnet or charter schools could also be created to specialize

---

* These are skills like phoneme awareness, phonological processing, semantics, syntax, and morphology.

in dyslexia-friendly education. These schools would not be extensions of special ed—where students with all sorts of challenges are clustered due to their shared inability to flourish in the regular classroom—but a gathering of students capable of flourishing in a specific environment that recognizes and nurtures their strengths while also training and accommodating late-blooming learners in areas like reading, writing, and other kinds of basic skills. These schools would be places where the feelings of stigma and failure that characterize the early years of education for too many dyslexic children need not be encountered. And they would be places where the sense of desperation to get students "up to speed" by age nine need not cast a shadow, because an expectation of developmental variability was built into the design.

Finally, a brief word of encouragement to teachers and homeschooling parents who are on the front lines of preparing dyslexic students for their futures.

We have often heard from dyslexic individuals how important their relationships were with each of their teachers in determining how well they learned. International design and branding star Gil Gershoni, who describes his work as dyslexic design thinking, expressed this beautifully when he told us that the most important factor in determining whether he would experience success in a particular classroom was whether he "felt seen" by the teacher—that is to say, whether the teacher really saw him as someone with potential and the ability to learn. "I hear this when talking to many dyslexic people around the world, that at some point in their life it was really important that somebody really *saw* them. It is somewhere between an emotional connection and a way of just listening and knowing how to acknowledge success—not as just like, 'I'm proud of you,' but specifically what it is that I see in you that is meaningful—so that you can understand it yourself and be proud in the way you march forward.

And that provides the building blocks for self-confidence in learning."

Assistive technology pioneer and brilliant dyslexic thinker Don Johnston shared with us a similar story from his high school years, where the experience of having a teacher who took him seriously as a student and expected him to do the same completely revolutionized his self-concept. And Jack Laws told us about his own revolutionary experience when he found two teachers in high school who were able to look beyond his difficulties with spelling and grammar to recognize the value of his ideas. This high sensitivity to the perception of others fits well with recent research showing that dyslexic students show enhanced emotional reactivity to emotionally charged scenes in videos.

Of course, we can't always be in a perfectly supportive environment—nor, according to Gil Gershoni, would it be an unmixed blessing if we could. According to Gil, "I think I also needed to have teachers at some points who could *not* see me. I needed them for contrast, because if you are loved everywhere, then it can create a kind of ignorance about your need to work to understand and be understood by different types of people who don't really see a part of themselves in you. So I think that contrast is a part of the recipe, and I would never really choose to change that. I think young persons need to have that kind of challenge at times, because if you never climbed the mountain, you don't know what it is to earn the view."

Learning to Learn with Dyslexia: Gary McGregor

Most of the successful dyslexic adults that we've profiled in this book can cite parents, teachers, or other adult mentors who have played a crucial role in their success. But far too many dyslexic students lack such a mentor figure. They slip through the cracks without anyone ever recognizing their talents or helping them figure out how to learn with a dyslexic mind. As a result, they leave school at best unprepared and at worst deeply scarred, with a sense of themselves as flawed and incapable of achieving their dreams.

It is all too easy to write off those who've slipped through the cracks as having had low potential—or as having possessed a different kind or quality of dyslexic mind from those individuals we've profiled. For those who feel tempted to think this way, we offer the story of Gary McGregor. As you read Gary's story, we invite you to wonder along with us about just how many other Garys there are who are slipping each day through the cracks in our educational system.

"I'm the first male in my extended family ever to go to college, so it was never really expected of me. My dad was a bricklayer and industrial multi-craftsman, and after my parents divorced, my mom cleaned homes and worked as a teacher's assistant for life-skills kids.

"School was just awful for me. I read so slowly and I couldn't spell to save my life—and the grammar . . . I just couldn't show anything I knew in writing. Math was a little better, but they always wanted me just to copy it their way, which I had no interest in doing. Basically, I never got the 40,000-foot view of anything, and for me to learn, I really needed to understand things on a higher level. So I didn't even make it past the tenth grade.

"At home I was in my own head a lot of the time, and I had a very active imagination, so I'd go in my room and make mock battles with LEGOs, or build little vehicles to run down the stairs so I could see what broke and then put them back together. I took everything I could find apart and I would build my own toys.

I also watched way too much TV and I thought I wanted to be Rambo, or Arnold Schwarzenegger. But I never really learned how to learn. My family weren't really into teaching, even if they were good at something, so there wasn't anyone to take me under their wing, or say, 'Let me show you how to do this.'

"After I dropped out of high school, I joined the army, but after a couple of years I got a medical discharge because I had severe allergic reactions to the preservatives they used in the food and I was sick all the time. That kind of crushed my dreams, because I really wanted to be a career soldier. And after that I just kind of floated around doing different odd jobs. I worked as a rigger and a deckhand on a tugboat, and I worked at a golf course fixing golf carts. Finally I decided to try what my father did, which was industrial craftsmen work. It paid really well for someone who didn't have a high school degree, and I would do anything from demolition to welding, laying brick, boilermaker work, petrochemical or chemical manufacturing and refineries—we were kind of a one-stop shop. We called it turnaround work, and it was something I was finally good at, and my family was proud of me. Although there were routine aspects to the job, most days you didn't know what you'd be doing, so there was excitement in getting to learn and experience all these new processes.

"That work really sparked something in my thinking, because I found that I was really good at imagining things in my head in 3-D. It was like in the movie *Iron Man*, where Tony Stark manipulates everything on a computer and can swivel it around. Once I had physically experienced things, I could do that in my head—mentally pull them apart, put them back together, and figure out what I needed to do, right there on the job site.

"I also had a natural curiosity and I liked to see how everything worked. I loved to learn about the processes, like refining crude—breaking it down into the different fuels and all the by-products and chemicals and seeing how they're separated. It was just so neat to see this kind of clockwork system producing things we use every day. I was always trying to figure these things out and asking all these questions and basically driving my bosses crazy. But it gave me a kind of playground to start practicing this visualization, and I just started putting it all together.

"Over time I came to realize there was an even better job than the one I had. Whenever we had a problem and we needed to build something new they'd say, 'We need to call the engineer.' So I was like, 'This guy must be important. Maybe I want to be an engineer. I like building things, and I hear they make pretty good money.' By this point I was twenty-two, and there were some other things going on in my life, too, that had me thinking that maybe it was time for a change. I had sort of hit rock bottom in my personal life. I was a black sheep who partied pretty hard and generally messed up. I knew maybe fifty words, and most of them were curses, but I could use them all as nouns and adjectives and conjunctions and every other thing, so basically I was like something from the *Jerry Springer* show. But I was starting to feel like maybe there were some things I had gotten out of my system and some emotional maturing that had taken place—and maybe mentally, too—and I was starting to do a lot of soul-searching and praying. And I had a buddy who kept saying to me, 'You gotta go to college,' and I was like 'Dude, you know I partied pretty hard. Like, I'm pretty sure my brain is fried.' But he just kept bothering me, so finally I said, 'Fine, I'll go take the stupid junior college test just so you'll leave me alone.' So I took the test, and let's just say the results were not encouraging. They found that mathematically I could add and subtract whole numbers, but I couldn't add and subtract fractions. They said they would put me into the second-lowest remedial class, and I'd need to take a full year of math before even starting algebra. Same for writing: basic remedial.

"So, I came back and said, 'It's not looking so good,' and I was on the verge of not wanting to even try. But my buddy kept prodding me and I kept praying, and I just had this really strong feeling like, 'Hey, I gotta give this a shot.' So I told my family, 'Hey, I'm going to quit my job and go to college,' and they totally freaked out on me. They're like, 'You're an idiot.' My mom cried, and it was not tears of joy. It wasn't any better at work. They said, 'You're a good worker, but you're not what we'd call the intellectual horsepower of the group. You'll be back in six months.' And I was like, 'Man, this is terrible, but I just don't care,' so I decided to give it a shot anyway.

"Well, I got to junior college, and amazingly everything just clicked, because they encouraged alternative thinking. In my earlier schools the teachers were always 'my way or the highway.' But now my math professor was very open-minded, and he said, 'Just do it however you want to. There are lots of ways in mathematics to get to the right answer—do whatever you're most comfortable doing, and just show your steps.' That was perfect for me, because I realized I could understand a lot of math equations visually and use the same kinds of mental simulations I did at work to understand math, and that worked really well. Previously I was always taught to learn math equations by rote, so I never realized that an algebra equation could have a physical meaning to it or be understood by real-world examples. It was always like, 'Put this in, and spit it out.' But I needed to build a construct in my mind so I could physically understand it before I could really learn it and be able to apply it. And I had to figure all that out on my own. But once I did, I was like, 'Man, I get this, I understand it.'

"Now I was learning math at a rapid pace, and before I knew it, after two years in junior college I was already well into calculus, and I actually got scholarships to go to a number of universities to study math. So I went from a high school dropout to graduating with highest honors from my junior college. But I still wanted to be an engineer, so I started looking at engineering schools, and I found out that the University of Texas, Austin, was one of the top schools in the country, and it wasn't far from Houston, where I was living, so I decided to go there. So I loaded everything I owned into my truck and moved to Austin.

"Honestly, it was hard for me at first because of the trauma I sustained as a kid, from being labeled dumb and stupid and slow and all these other negative words. I mean, those words can really hurt, and while physical pain can go away, some of that mental trauma, especially in the earlier developmental phases, can last straight into adulthood, and then linger there for the rest of your life if it's not addressed. So one of the things I had to learn was about self-talk—what you're telling yourself and how you're viewing who you are. Whenever something would come up—whether it was a problem, or even when I succeeded and got a good grade on a test—I would just sort of freeze up, and self-doubt would creep into my head, and I would feel like I'm faking it, or that this is just a lie, and it's all going to come down on me at some point. And that was something I just had to get over. I worked hard to reverse my inner dialogue so that, for example, when I would see some crazy math problem, instead of panicking and saying, 'What do I do?' I'd be like, 'Wow, this is interesting. I wonder what this actually does. Let's be curious and playful with it.' And I'd approach things from an attitude of curiosity instead of worrying like, 'I'm going to fail.' And I think that any dyslexic student who has gone through this kind of trauma needs to also undergo this kind of mental rewiring in their self-talk, as well as learning about how to learn in the purely academic sense.

"There was plenty of this latter kind of work to do, too. I felt that everyone else had one up on me, because they all had teachers or parents who'd pushed them to study and taught them how to, while I was having to learn all that on my own. So I had to find students that were really good at studying and ask them lots of questions and just experiment with myself to see how I could learn.

"Writing papers was the biggest problem for me, so I'd always try to find the oldest person in the class who wasn't me and try to make friends with them so they'd take time to help me proof my papers. At first my spelling was so bad even spell-check wouldn't help me, and my grammar . . . But the more you do it, the better you're going to get. So I improved dramatically, especially when I would take their feedback and look at their corrections. But even now it's so embarrassing for me to get up in front of a dry-erase board and start writing, because I'll reverse letters in words or use the wrong pronoun or whatever. Like the other day when I spelled *testosterone* like Testarossa, which is a kind of Ferrari, so I'll usually pass on the whiteboard.

"Basically, when you're dyslexic you're also going to take longer on homework and longer on tests, and you're going to have to struggle to learn to spell words and things. But you're just going to have to deal with that. On the other side, you're also really going to be able to understand the material well if you can learn to engage your strengths, and for me those were visualization and stories. I found that if I could visualize something and get it in my head in the form of a story or some context where I could visualize it, then I could manipulate it and do whatever I wanted with it, and I could gain mastery over the material pretty quickly. For example, I remember a material science course where all the students were struggling to grasp the concept of how the size of the atoms in a material affected diffusion and hardness. And I could literally just start drawing analogies in my head—like imagining a jar full of marbles, and then putting sand in there—and then imagining different sizes of marbles and how the sand would flow through there. And that made it really click for me, while everyone else really struggled. And I really understood that material. It's also stuck with me well, whereas others might just have learned it to pass the test. But in engineering and science, all those things you learn tend to be building blocks for something more advanced, so a deeper understanding that lasts beyond the test is really important.

"I also needed to learn how my memory worked. I have an atrocious rote memory, and if you just flashed a series of numbers in front of me, I couldn't remember more than five. However, if I can turn it into a story or into some image I can visualize, I've been able to memorize entire tables of engineering figures, and people think I have a really good memory.

"I've also learned that a really good way for me to solidify my learning is to talk to someone who's a real expert. If you've already done your homework, and you pre-studied the field and you understand it so you know the right questions to ask and you're not asking them to teach you from ground zero but just to clear up some confusions, I've found that people really respond well to that, so I do that a lot. I've also found that most teachers have particular strengths and weaknesses in how they explain things, so I would often go out and buy multiple different textbooks on a subject, and then, because the different authors explained the same things in different ways, it would help things click so I would really understand them.

"And with all these strategies, I was able to do well enough to get into the UT PhD program, where I earned my doctorate in biomedical engineering. But it would be almost a crime to call me a biomedical engineer because I'm really a general device person who likes to build and develop all sorts of things, from thermal fluid systems to medical tools to surgical simulators. Currently I'm vice president of operations and principal mechanical engineer for an engineering firm and a cofounder of a medical device company.

"And to be honest, one of the things I've benefited most from both in school and in my engineering work is that I did those jobs in heavy construction and heavy industrial systems, because that experience has made things a lot more intuitive and easier. When you have students who've been purely educated in a school classroom and you give them complex problems that they can't easily visualize because they've never physically seen anything like it, that can be tough. In my materials class when they were teaching us about welding, I'd been a welder, so things just stuck for me in a much more meaningful way.

"I'm sort of a natural polymath, who can pick up a lot of different areas of knowledge or skills from just bouncing around. The cost of that approach is that you can lose focus or not be able to dive deeper. There's also a time cost, so you have to be self-disciplined and not get pulled in too many directions. But that natural curiosity and wandering outside of boundaries is also what contributes to a lot of inventions and thinking outside the box. So, for instance, when I was working as an engineer for a while in a steel mill and they were having a certain problem, the fact that I worked in refineries and understood their systems allowed me to adapt a technology from the refinery to work in the steel industry. So crossing over disciplines has benefits, but you have to be careful to not stretch yourself too thin.

"I'd say the same thing to parents of dyslexic kids: Let them be exposed to as many experiences as possible. The more a dyslexic kid gets exposed to, the more things they have to draw upon for their nonlinear thinking. So feed that curiosity about how things work, and be patient with your kids if they're asking questions like, 'Why is this?' or 'How does this work?' Let them do things and let them make their mistakes or destroy a toaster or take apart the coffeemaker—but safely! I also think parents tend to worry too much about grades in the early years and not enough about feeding that curiosity. At Georgia Tech, which is one of the top engineering departments in America, they did a study to look at which graduate students succeeded in their research projects, and there was essentially no correlation with grades or with standardized test scores on the GRE. What really mattered was tenacity and creativity. Tenacity, because you have to learn how to deal with failure either in research or in engineering in general—and I've really seen that firsthand with some of my peers. If you've gone through your entire K through 12 education making perfect scores and you're so used to success, you can have a hard time dealing with failure, which is an essential part of research and development. I've also seen some of my peers become frustrated and drop out because they couldn't take that. But in research you want to fail early and often, so you can get all the problems out of the way. And to deal with that failure, you need tenacity. And because dyslexic students are used to the failures, there's no problem there. We can just move past that and reset and get over it. And the creativity is also something that dyslexic students also seem to be good at—making bigger leaps and doing things new ways. I really think there should be more dyslexic people in science and engineering."

Gary's story provides a particularly clear example of the truly remarkable changes that can occur when dyslexic thinkers *learn how to learn* by using and building upon their strengths. Our experience has convinced us that there are many more Garys among our struggling students than almost anyone imagines, and they have a similar potential for radical transformation. And certainly dyslexic students will learn how to learn better if we simply take the time to show them how to understand and leverage their strengths.

33

Dyslexic MINDs at Work

In this chapter, we'll focus on steps that both dyslexic workers and their employers can take to gain maximum advantage from the benefits provided by the MIND strengths. As we have just seen both through Gary McGregor's story and the chapter on complementary cognition that preceded it, there is a phenomenal reservoir of untapped talent just waiting to be unleashed by employers with enough vision, daring, and energy to select a cognitively diverse workforce and create environments to facilitate their synergy. We stand on the threshold of a workplace revolution, and in this chapter we'll provide some ideas on how to step across that threshold and enter an exciting new world.

Putting the MIND Strengths to Work

Before looking at each MIND strength in detail, remember that all the MIND strengths are built on a set of common systems, so many dyslexic individuals will show strengths in several MIND areas. Nevertheless, because the MIND strengths are all *modular functions*—that is, they arise from combining these common systems with additional factors—each individual's MIND strengths may vary in expression. For example, some dyslexic individuals with tremendous 3-D spatial reasoning abilities may struggle to remember nonphysical aspects of their past experiences, which is a component of N-strengths. Others may reason very effectively about the spatial properties of objects but have a poor sense of their own location in space. Still other dyslexic individuals may show poor spatial reasoning but excel in systems-based reasoning (I-strengths). There are as many variations as there are dyslexic individuals.

Also, given the modular nature of mental functions, many factors that lie outside the MIND strengths—including language, quantitative reasoning, attention, interests, motivations, previous experiences, and even beliefs—can both modify the strengths and have a great effect on the intensity and effectiveness with which any given individual may apply them. As a result, MIND strengths will reveal themselves in different ways in different individuals. So when considering the implications of a given individual's MIND strengths for work, remember: "Your mileage may vary." Treat the various recommendations and examples we provide below as helpful hints, to be tested and then either accepted or discarded based on their usefulness, rather than as hard-and-fast rules to be rigidly followed.

M-Strengths

Dyslexic people with powerful M-strengths are a tremendous asset for any form of work that requires thinking about physical objects or spaces. Many of these activities will be fairly obvious (for example, the mechanical and building trades), but others may not come so readily to mind (for example, orthodontics or art). For those who are interested, we have included a list of jobs that especially engage each of the MIND strengths on our website at dyslexicadvantage.org/mind-careers/.

Two of the key applications for M-strengths in the workplace are *creating plans* and *reviewing plans*. Not every dyslexic individual who shows prominent M-strengths will excel in both, so being aware of the difference is important.

Creating plans using M-strengths (for spatial objects, buildings, machines, layouts, activities, etc.) typically also involves the application of other MIND strengths—the creative abilities of I-strengths, the narrative and human-centered N-strengths powers, or the predictive force of D-strengths. When impressive spatial reasoning abilities couple with these other strengths, they can create a powerful inventor, as we saw in the case of Yoky Matsuoka, who used her interdisciplinary and analogical thinking strengths along with her spatial abilities to create highly functional robots, or a creative scientist, as we saw with Jack Horner, who combined his spatial abilities in detecting fossil patterns with strong mental simulation and predictive abilities to puzzle out the past conditions that must have led to their current state. The best plumbers, mechanics, surgeons, and artists are great problem-solvers as well as great technicians, and this creativity comes from combining spatial reasoning with other strengths.

Reviewing plans using M-strengths can also involve mental simulation and prediction, but it leans a bit more heavily on "pure" M-strengths than creating plans does. As a result, most high M-strength individuals will be good at it. Reviewing plans involves looking at a set of blueprints for a packing arrangement; constructing an object, building, or machine; or coming up with a layout for an office, room, or factory, and then answering questions like: Will it function? Will it fit? How will it look? What will be the hidden impacts? Will there be indirect consequences? The ability to pick up the subtle flaws in plans before going through the time and expense needed to mock them up—or worse, build them—is invaluable for any business, and this is an area where dyslexic workers often prove they are tremendously valuable, as we have heard again and again from employers in construction, manufacturing, design, film production, and other high spatial fields.

Importantly, the tasks of creating and reviewing plans can often benefit from the involvement of several collaborators with complementary strengths. For example, an individual with very strong mental spatial simulation but weaker powers of prediction or divergent creativity would work even better with someone who specializes in those skills; and having a detail-oriented person to make sure that attention had been paid to all relevant issues and that the M-strengths "spotlight" is pointed in the right direction is also a plus.

M-strengths can play a valuable role at every level of a process that requires spatial reasoning, from the offices and cubicles where plans are created to the workshops, assembly facilities, or job sites where they are implemented. In high spatial fields, you will often find M-strength thinkers throughout organizations—including individuals near the top of the management hierarchy who began at the bottom and worked their way up, taking advantage of related skills at each level. For young dyslexic individuals who might eventually want to hold high-level jobs in one of these fields but aren't quite ready for additional schooling, spending a few years working in an entry-level capacity in that field before going on to pursue the additional school training might prove to be valuable, just as Gary McGregor has found his background in construction and mechanics extremely useful in his work as an engineer.

One important thing that employers of dyslexic individuals with prominent M-strengths should always keep mind is the often profound difference between the strengths these individuals show in spatial reasoning and their ability to convey their ideas about spatial concepts in the linear-sequential medium of words. Think back to earlier sections where we discussed both the trade-offs often seen between spatial ability and verbal fluency, and the nonverbal nature of the thinking reflected in the questions on the M-strengths survey. Employers can create unnecessary obstacles to success by requiring individuals with prominent spatial strengths to describe or justify their work in written reports or proposals or to pass written tests or qualifying examinations without appropriate accommodations for reading or time. These individuals must be judged by the quality of their work in their medium of strength, rather than on their ability to translate information from that medium into words. Likewise, dyslexic individuals with prominent M-strengths should learn to partner with or hire the appropriate verbal talent in situations where writing demands exceed their abilities in certain high-stakes areas, like creating a grant proposal, submitting an application, or writing papers or theses.

I-Strengths

Individuals with I-strengths are valuable for almost any kind of work, but we will focus on three key contributions that they can bring to the workplace.

We'll call the first contribution *fresh thinking*. It covers a range of innovative abilities from divergent creative thinking to reframing problems by paring them down to their essence or gist to disrespecting boundaries through interdisciplinary approaches—in general, proposing novel approaches to old or existing problems. Often a dyslexic mind will provide the dynamite that breaks up the logjams caused by an inability to step outside old assumptions. Because dyslexic individuals with high I-strengths tend to approach problems from different perspectives and use techniques that differ from those of other people, they can often find ways around problems that seemed intractable to persons trapped within a set of assumptions and limited to current methods. Having a fresh-thinking high I-strength dyslexic person in literally any group that is tasked with innovation—whether in engineering and design, corporate management, human relations, or sales, marketing, and advertising—can help prevent this kind of blinkered thinking.

The second contribution is *systems reasoning,* or the ability to understand and reason about complex systems. Some dyslexic thinkers excel in understanding relatively stable systems like factory assembly lines, which are complex in their formation but function with little variation. Others (who combine strong D-strengths) excel at dealing with more dynamic or changeable systems, like farms or complex supply chains, where elements like weather and disease can be managed but not controlled. It is important for each dyslexic person to know what they do best. The applications of strengths in systems-based reasoning are essentially endless, and everywhere you find such complex systems operating successfully, you will find a high incidence of dyslexic thinkers, particularly in the planning and troubleshooting phases (that is to say, "What's wrong?" and "How can we fix it?"). Partnering with expert exploitation- and detail-oriented thinkers for day-to-day operations can create an ideal mix.

The third contribution is *facilitating teamwork,* or helping groups of people function better together. This doesn't always mean being team manager. It may just involve providing insights into the ways that a group of people can collaborate more effectively to aid the group's manager in her or his work. It's really just another form of systems thinking, but in this case the system is a human one. In complex systems like large organizations and bureaucracies, high I-strengths thinkers can show special skills in knowing how to work the system or get around barriers. They may also excel in connecting people within the system who have different objectives and priorities, work in different departments, or even speak in differing specialized jargons, like engineers and marketers. Any activity that involves detecting, understanding, and coordinating the skills and activities of groups of people—from teaching and coaching to office management to theater production to running a busy machine shop—would benefit from including a high I-strength individual.

Employers of high I-strength individuals should keep in mind a few potential challenges. First, high I-strength individuals can be extremely hard to accommodate within the demands of a rigid organizational chart. With their deeply interdisciplinary minds and their lack of perception—or fear—of traditional boundaries, it can be difficult to know where to put them, how to delimit their jobs, and how to keep them within those boundaries. They will benefit from mixing with colleagues in different areas or divisions, and also from being part of a more lively, playful, and interactive work environment. Staying as flexible as possible is a good rule of thumb, though occasionally they may need refocusing. Many highly creative people are subject to the lure of the "shiny penny," where each new idea seems like the most attractive, so maintaining at least some balance between exploration and exploitation is as important at the individual level as it is at the group level. But in general, remember that these are good problems to have.

Second, they will get their work done in ways quite different from those employed by their colleagues. Focusing on results rather than methods will be essential.

This advice is also useful for dyslexic workers themselves. Because of the way your mind works, you may often be drawn to jobs or projects that lie on the boundaries between two or more domains. Be aware that some people in the relevant departments, professions, or fields may view you as an outsider or even a threat. Building good relationships with those individuals will be critical not only to prevent pushback but to maximize your opportunities for fruitful collaboration. Focus on how you can create win-win situations, and learn to communicate the value that working with you will bring to others. If you're not good at this already, receiving coaching from someone skilled in this type of communication may be helpful.

N-Strengths

N-strengths also have many practical applications in the workplace. We'll again focus on three. The first application is *storytelling.* This is the ability to use personal strengths in episodic construction and mental simulation to create new narratives that can persuade, convince, explain, or educate, as well as to entertain and engage in rethinking old conclusions. Essentially any job that involves human communication can benefit from these skills. Rather than persuading through logic, individuals with N-strengths build new models, then invite others to enter them and experience a new perspective. Sometimes this is a simple act of reframing: of taking familiar events and experiences that a client has pointed up and reassembling them in new arrangements that open up fresh and better interpretations. These skills have countless uses, but one of the most important is the story dyslexic workers learn to tell about themselves. Often, before individuals with dyslexia can show what they can accomplish, they must convince others to give them that chance. This requires self-advocacy: the ability to say, "Here's what I do well, and here's how I think I can best contribute." Self-advocacy involves persuasion and negotiation, so learn to use these convincing storytelling skills to your advantage.

The second application we'll call *personal touch*. As we've discussed, dyslexic individuals tend to store more memories as specific events rather than as generalized schematic memories. As a result, they are more likely to remember particular cases from previous experience, especially unusual or outlier ones. This collection of rough data equips them to spot potential real-world problems when planning or troubleshooting that might be missed using generalized or aggregated data. This bias toward the particular can also be useful in any task that requires remembering the preferences or particular characteristics of individual clients in jobs involving sales, customer service, counseling, and teaching—but remember, personal memory strategies may need to be used to store and retrieve these details.

The third application is *learning from experience*. Dyslexic people with prominent N-strengths are often especially good at picking up information—both intentionally and incidentally—when working with clients or making site visits. As such, they are particularly strong in fields like user design, customer service, sales and marketing, or product testing, where learning from customer feedback and experience is essential. Their broader field of attention can help them detect not only the things that others intend to show them but also more obscure details that others might be trying to hide, and can enable them to see how those extra bits of information fit together to create a story that provides clues regarding unspoken motives or goals. So it is always good to have a high N-strength dyslexic person involved in visits, inspections, and negotiations.

Employers will do well to remember how much better many dyslexic individuals learn through experience than through seminars or other types of formal instruction. Often a recently hired dyslexic employee who appears to be just right for the position may show difficulty absorbing the new information essential for doing that job through a lecture-based training session, yet that employee may learn quickly and well via on-the-job training or through apprentice-type relationships with other workers. Dyslexic employees should not be assessed through written tests, but orally or through observation of job performance.

One other thing to keep in mind is that high N-strength individuals often thrive in situations that involve human interaction, or at least directly impact other human beings, rather than more abstract or theoretical work. Making sure that these employees have a firm sense of the human meaning and importance of what they do is critical.

D-Strengths

D-strengths also have three extremely valuable work applications. The first we'll call *triage,* which is the ability to sort through a series of simultaneous demands quickly and accurately to decide where to apply available resources to achieve the most positive outcome. This kind of weighing and balancing multiple competing demands and priorities is exactly the kind of complex multidimensional modeling at which many dyslexic people excel, and which is almost impossible to perform well in a timely fashion using more standard logical-deductive approaches. Triage skills can be useful in any busy workplace, and the more demanding the environment, the more useful the skill (e.g., an emergency room, an air traffic control tower, a busy construction site, or a restaurant kitchen).

The second application is also useful in dynamic settings, and we'll call it *Plan B*. Plan B is the ability to respond on the fly with a best-fit solution to new and changing conditions when information is incomplete and time for research and reflection is simply not available. This is another situation where using experience-based mental simulation and the default mode network problem-solving is essential, and step-by-step rule-based approaches fall short.

The third—and in many ways, the most characteristic—D-strength application is *prediction*. Prediction requires both a highly accurate mental model and the ability to simulate changes that will occur in that model over time due to the operation of various processes. Both forecasting, or predicting what is likely to happen in the future, and backcasting, or retrospectively figuring out what must have happened in the past to lead to the current circumstances, are extremely useful for many kinds of work. Strong dyslexic predictors can form powerful partnerships with analytical problem-solvers who can help them work backward from their predictions, so those predictions can more clearly be explained to others with a more step-by-step thinking style. In our work we have encountered many examples of highly successful simulators who have partnered successfully with step-by-step thinkers to create teams of amazing power—as, for example, when visionary CEOs with a clear sense of where to take the company combine with detail-oriented COOs who can implement the plan, or investors who excel at spotting trends work hand in hand with skilled financial analysts who can affix numbers to those expectations.

Employers of high D-strength dyslexic individuals should also be aware of the frequently paradoxical nature of what these employees find hard and easy. Typically, they find jobs that involve repetition of routine tasks, attention to detail, and use of standard procedures and fixed rules to be extremely difficult, yet they may thrive in environments that contain novelty, change, and challenge. Ida Benedetto, an experience design specialist and highly creative dyslexic thinker, expressed this apparent paradox to us in this way: "I recently spoke with a colleague who told me about her previous job. She said everything on that job was so certain and clear. There were no human or personal factors involved, and the work was highly procedural and very straightforward. 'It was just so easy,' she said. 'It was the perfect job.' And I told her, 'I think you need to use an "I" statement when you call that job "perfect" and "easy," because that's a place where I would totally suffer.'" Ida told us that the perfect work environment for her is to "deal with leadership teams at major corporations, where I come into new situations and diagnose what's going on, then help them either envision a new purpose for their organization or come up with new ways of working that are more effective for responding to highly dynamic, ambiguous, and adaptive environments, by building new narratives." As a result of similar environmental preferences, many dyslexic entrepreneurs and tech professionals have a fondness for working in early stage companies or on new initiatives in existing companies, where each day brings something unique, job descriptions and assignments are very fluid, and there are plenty of problems to be solved. Once processes become established and the days become more routine and unvarying, these early stage specialists may head off to find a new venture.

Conclusion

Whenever we think about the vast untapped pool of talented dyslexic individuals that lies waiting to be discovered by forward-thinking employers, it reminds us of an analogous situation that existed almost thirty years ago in professional baseball, as told in the book *Moneyball* by Michael Lewis. Tasked by the owners of the Oakland Athletics baseball club to assemble a winning team while spending less for players than anyone else in baseball, the team's general managers used new forms of statistical analysis to uncover hidden value in players nobody else wanted. Eventually they created teams that came within a whisker of winning the league championship on just a fraction of the payroll of most of their competitors, and in the process revolutionized the way talent is assessed in baseball players. What was their secret? They discovered that the traditional metrics being used by baseball's talent scouts to identify promising players were simply measuring the wrong things, and that many players who could contribute greatly to team success were being passed over simply because they didn't *look* like what everyone assumed a star player should be.

Currently, many employers are stuck right where baseball scouts were thirty years ago. They are overlooking a massive pool of talent because it doesn't resemble the traditional picture of a talented worker they've inherited from the past: excellent grades from top schools, high marks on aptitude and certification tests, and perfect spelling and mechanics on emails and applications. The key to benefiting from the immense reserves of creativity and problem-solving strengths that this talent pool contains is to throw out old prejudices about what it means to be smart and start looking for the right sorts of things—like the kinds of strengths discussed in this book.

Working with Dyslexia: Chris Ford

We'll end this chapter with a story about a remarkable young man whom we first met almost a decade ago, shortly after the first edition of this book came out. At that time, just like many of the players selected for the Oakland Athletics, he had gone "undrafted" after finishing medical school, failing to match for a residency position at any of the training programs in the United States in his chosen area of specialty. And just like many of those players, he has since become a star.

"I was born in Chicago, where my mom was a teacher and my father was a firefighter. They always emphasized to my brother and me that education was our key to succeeding in life. Not just by their words, but by their actions. My father started out at the lowest level of the Chicago fire department when it was still very segregated, and the opportunities were limited for African Americans, but he was very ambitious and determined to break those barriers. I remember that he took the lieutenants exam six times, and each time they said they 'lost' his answer sheet—and remarkably those of the other minority examinees as well. Eventually, the FBI investigated and took the examination process away from the fire department, and that's when the talented minority candidates started to rise. But even when the system was corrupt, my father never stopped fighting. I remember I'd come home from school or football practice and he would be sitting over books with six other firefighters, holding study groups at our house, and he would teach these other people how to study for the exams. They would work hard, and they would score very high on those exams, and a lot of those folks eventually became leaders in the department. When my father retired last year, he had risen all the way to be fire commissioner in Chicago. I'm very proud of my father, and I've learned from his example.

"I went to Chicago public schools on the South Side. I had a lot of trouble learning to read, but spelling was even harder. I remember getting chastised by my teachers: 'You're obviously smart, but you need to start paying more attention to your spelling!' But at that time, they didn't figure out that I was dyslexic, and I didn't receive any specific instruction.

"It took me a long time to figure out that with reading, one of my big problems was that I was always losing my place. I'd get to the end of a line, then skip the next line or a whole paragraph. Math was also a problem—not because I couldn't get the right answers, but because I couldn't do things the way I was taught. I'd come up with my own methods, and some of my teachers were cool with that, but others were not. But my mom said, 'As long as you get it right and can move on to the next step, that's what's important, not your grade.'

"In high school I was an athlete, and I did the bare minimum to get through, so I graduated with about a 2.5 on a 5-point scale. I got into college on a football scholarship. Then suddenly all my parents' messages about the importance of education for the rest of my life finally took hold. I chose to major in biology because I thought I might want to be a marine biologist, but I noticed that it took me much longer than most other people in my classes to learn. I still didn't know anything about dyslexia. I just knew that if I read the book, it took forever and I'd get maybe 10 to 15 percent of it the first time. Then I'd have to go back over it again and again. But my attitude was, *I just have to work harder.*

"In my third year I arranged a course where I shadowed a

plastic surgeon around in the hospital. I saw his connection with his patients and the good he was able to do, and I decided that I wanted to become a doctor. That made me work even harder, and I essentially lived in the library during my undergrad years, but I was able to earn a 3.75 science GPA when I graduated.

"Then I got into medical school, and it was suddenly like trying to drink from a firehose. I got an apartment right across the street from the medical college, but the cubicles in what we lovingly called 'nerd room' were my real home. I'd get there at eight o'clock in the morning, study, take breaks only to go to classes or eat or go to the gym, then finally go to my apartment [at] about two in the morning to take a shower and get a couple of hours of sleep. Then wake, rinse, wash, repeat. While that had worked well enough in undergrad, it didn't really work in medical school, and I noticed that the work I was putting in was not showing up in my test scores. I met a classmate who's now my wife, and she would wake up in the morning and go to a coffee shop and study her notes for a while, then go out and live an actual life, and she was doing way better than I was. I thought, *How are you doing this? You must be the smartest person in the world.* That forced me to realize there was something about my learning process that wasn't clicking.

"After the first year we had a big exam called the Step 1 test that looked at everything we had learned to that point, and although I passed, I didn't do nearly as well as I thought I should, given how hard I'd studied. I met with one of the deans at the medical school and told him about my frustrations. He had them perform some tests on me, and that's when I was diagnosed with dyslexia. At that point neither of us really had a clue what that meant or what role it was really playing in my challenges, but we were both sort of curious, so we started a process together where we tried to piece out how I learned and what I could do to learn better. He gave me some more tests to see how I took in information, and I also did some searching around on my own. That's when I found your book, The Dyslexic Advantage, and reached out to you. I saw myself in each of the stories in your book, and I slowly found ways that helped me learn better than the brute force read-it-and-memorize-it approach I'd been using. Since that time, I've been able to use those abilities to my advantage on a daily basis.

"I found video resources online that worked much better for me than reading, and I also started to take a more conceptual approach to learning, trying to really understand the information rather than just memorize. I remember that my neuroanatomy class was the first where I really tried to do this—to visualize and understand—and I ended up getting 100 percent on that test. By the time the second Step test rolled around, I was able to score about fifty or sixty points higher, and that much higher still on the third Step.

"Despite the fact that my grades began picking up, when it was time to apply for residency training, I applied in emergency medicine, and there were more applicants than training positions available. I only received two interviews and wasn't offered a position anywhere. I felt devastated, and that was definitely the low point of my life. My girlfriend (now wife) matched to a pediatrics residency program at the University of Minnesota, so I moved there with her, and I began working two jobs. The first was as an internship at the university, where I examined and enrolled patients in clinical research studies; and the second was a nine-to-five job working for Epic Systems, the big electronic medical records firm. In my spare time I also begin drawing up blueprints for medical devices.

"I began that year pretty depressed, but that was actually a great time for me, because everything I did connected together and started building in positive ways. I became very good at coding in HTML, and I learned to create macros and dot phrases for streamlining and simplifying the process of medical record keeping. I also got to travel all around the Midwest to help hospitals set up their medical records systems, and the skills I acquired that year have been a huge help to me ever since.

"After that year I applied again for emergency medicine residency programs, and this time I was invited for fourteen interviews. I was very up front about my dyslexia in my application because I wanted to make sure that I chose a program that wouldn't see that as a problem. In about twelve of those fourteen interviews, I talked with the program directors about my dyslexia and what that would mean. In general, they were very receptive. Most just wanted to know what it meant, because even though we were in a medical field, most did not know how it would impact my performance as part of an interdisciplinary team in emergency medicine. But they were also impressed by my research and my work with Epic, and at the University of Wisconsin, one of my interviewers was also very interested in my medical device plans, because he was creating devices, too.

"I ended up matching at Wisconsin, and they were a fantastic fit for me. The faculty were very flexible in their thinking, and they allowed me to do extra work in pediatrics and medical device development. And the professor who showed interest in my devices helped me develop one of them for production, so it's now in use to help doctors learn to intubate patients more safely. The faculty were also very receptive of my needs as a dyslexic person, and every day I'd get an email from one of the people on our staff or from my program director asking how things were going or if I needed any accommodations. They also asked frequently if I had anything to bring up to the educational committee at the medical school, and eventually they asked me to become part of it. One standard practice in medical education that I was very up front about saying I would not benefit from was what's called *pimping,* or the kind of oral quizzing that requires you to rapidly retrieve and verbalize factual answers on the spot when an older resident or attending asks you a question—often right at the patient's bedside. That's not something I can do. But if we sit down and talk about things, I can clearly show that I understand what's going on. So even though they weren't initially very knowledgeable about dyslexia, they really wanted to learn how we could make it into something positive.

"Another bane of emergency medicine is the amount of documentation, but assistive technology helps a ton. I developed systems to help with data entry and retrieval for laboratory results and medication orders, and worked closely with our pharmacists and technicians, and I learned to use technologies like Nuance Dragon NaturallySpeaking to create patient records almost entirely orally, often while I'm just walking between different patient rooms or beds in the ER.

"Residency was also a great time of learning for me because I'm a visual and auditory learner, and I learn best by doing. I had great clinical training, and I also discovered a fantastic podcast called *EM:RAP*, which goes over important articles in emergency medicine and covers information about procedures and medicines and how to avoid mistakes. That podcast format worked much better for me than books, because I could listen whenever I worked out or had downtime. I also used another great video program and took notes on everything using a program called Evernote. So from the very first day of my residency, I was already essentially preparing for my board exams. As a result, I became really knowledgeable in my field and also really good at teaching the younger trainees and medical students. So in my third year I was named one of the program's three chief residents and was placed in charge of medical education for the entire residency program. The other two chiefs also focused on their best areas of competency—administration and scheduling for one, and resident wellness for the other—and together we won the award that year as the best team of emergency medicine chief residents in the United States.

"In the end, emergency medicine has turned out to be a really great fit for me because many of the skills involved—like managing a team to perform different duties on complex tasks, or anticipating and visualizing what's about to happen in these rapidly changing situations—are things I naturally seem to do well. We were taught to ask, 'What is the rate limiting step for each patient? What's keeping them here, and what will tell you either that it's okay to discharge this patient safely to home or else to admit them to the hospital?' Answering those questions requires running a lot of alternative scenarios through your mind, and I've found that I've been able to do that more easily than many of my colleagues.

"I've also noticed that I think about medical procedures differently than many other doctors. When we were being taught to do new procedures—for example, intubating a patient, sticking a breathing tube into their windpipe so they could be artificially ventilated—unlike most of my colleagues, I wasn't just trying to learn how to do it. I was also noticing all the different ways things could go wrong—both from the practitioner's perspective and the patient's—and thinking about how you could keep people from making those mistakes. That's where a lot of my medical device ideas have come from. I've continued to work with students in the Biomedical Engineering Department at the University of Wisconsin on medical device ideas, and in total I've developed eight different prototype devices, mostly for educational applications, and several of them have been licensed for production.

"I've also found that my ability to look from different perspectives has been very valuable in a number of leadership roles. When I was in medical school, I was selected by my classmates for the Association of American Medical Colleges' Organization of Student Representatives. I attended biannual national meetings where we would bring and discuss concerns and ideas from our medical schools, interact with people who lobbied for legislation, and then take information back. Many of the people I met at those meetings now hold important political positions, and I'm still interacting with them around areas of concern, like my work with disadvantaged children in the Milwaukee area.

"I was also selected last year by Wisconsin governor Tony Evers to join a commission of citizens from each state congressional district called the People's Maps Commission to create a nonpartisan redistricting plan, and my co-commissioners did me the honor of selecting me as chair commissioner. That's been a real learning experience as well. I went in knowing absolutely nothing about redistricting, but I decided I'd use the same audio and visual learning approach that I've used in learning emergency medicine to get up to speed. So I started off with a film called *Slay the Dragon,* which presented a big-picture perspective of the subject, then funneled down through audio lectures, videos, and doing hands-on work with some of the software programs that are used in redistricting. Now I can *see* it and I can *feel* it, and I know how I can make it better by applying what I've learned. I also think my spatial reasoning abilities have been very helpful in analyzing the census blocks and seeing how districts could be divided up.

"I'm also active in helping dyslexic medical students and residents, giving them pointers that I've found useful and kind of paying it forward. I think that's something really beneficial in the long run for all dyslexic people to do."

It's impossible to come away from speaking with Chris without a deep conviction that this is a young person destined for great things. Yet without his almost superhuman capacity for hard work, his eventual just-in-time discovery of ways that he could leverage his strengths, and the commitment of those in his medical school and residency training program to turn the experience of working with this brilliant young dyslexic thinker into something positive, the field of emergency medicine might have been deprived of one of its emerging leaders. Fortunately, Chris's story now provides us with a perfect example of the win-win benefits that everyone can enjoy when employers take even a few simple steps to support their dyslexic workers.

34

Modeling the Dyslexic Self

Throughout this book we've discussed ways to build multidimensional models of dyslexia and the dyslexic mind. In this chapter we'll ask every dyslexic reader to build another, broader model, which will include other features. This broader model will represent yourself as a whole—that is, your self-concept. In building this model, you must decide what role dyslexia will have in it.

Perhaps it seems presumptuous to assume that being dyslexic will form a significant part of the self-concept of every dyslexic individual. After all, dyslexia is just one part of each dyslexic person's makeup. Yet there are two reasons to believe that dyslexia will play an important role in forming the self-concept of most individuals with dyslexia—whether they embrace their dyslexic identity or not.

First, the cognitive differences underlying dyslexia create distinctive strengths, challenges, problem-solving styles, ways of learning, interests—even different patterns for development.

These differences impact so many of a dyslexic person's experiences at school and work that they will almost inevitably play a large role in shaping self-concept. That's why learning to understand these differences in a positive way is essential for a healthy self-concept.

Second, being dyslexic has already been shown to have enormous—and at present, too often negative—effects on the self-image of most dyslexic people. Dyslexic differences create classroom challenges for children at a time when school is their main occupation and academic success the main measure they use to compare themselves with classmates. Research has shown that these early struggles cause nearly universal—and often lifelong—effects on the self-esteem of dyslexic learners.

In short, if you are dyslexic, then *being dyslexic* will almost certainly affect your self-concept, so it will be best if you can learn to understand its role in your life in a healthy, well-informed, and productive way. In the sections that follow, we'll present some of the best ways we've seen to do this, which we've learned by observing the many thousands of dyslexic individuals with whom we've been privileged to work.

Embrace Your Dyslexic Differences Wholeheartedly

In our experience, individuals with dyslexia who embrace their dyslexic differences wholeheartedly generally have the most positive self-concept. This does not mean that they enjoy every aspect of being dyslexic—like problems with spelling or slow reading or difficulty filling out forms or taking notes. Far from it. But it does mean that they have a bigger vision that accepts these challenges as trade-offs well worth accepting for the sake of other strengths. They understand that their minds are optimized for different purposes, and they are comfortable with that fact.

Three things can help greatly in forming a positive vision of being dyslexic. The first is to learn as much as possible about your own thinking processes—both what works well and what doesn't. The formal term for this kind of thinking about thinking is *metacognition,* and a sound metacognitive understanding of your own thinking profile can help you tremendously in achieving success and self-understanding. Such a profile will include an understanding of some of the key points about dyslexic minds that we have shared in this book, such as their biases toward:

- Big-picture strengths versus fine-detail weaknesses
- Conscious and deliberate versus automatic processing for many basic skills
- Experience-based (episodic) versus abstract and generalized (semantic) long-term memory
- Diffuse versus focused attention
- Exploring new opportunities rather than exploiting those already known
- Learning by doing versus rote memorization and repetitive practice
- Reasoning and problem-solving using mental simulation versus more formal logical or deductive approaches

The degree to which these biases hold true for each dyslexic individual will vary somewhat, as will the presence of the various MIND strengths, so all dyslexic individuals must perform their own self-inventory. But these rules of thumb are a good place to start. So are the MIND strengths surveys included in each of the "What's New" chapters. Many dyslexic people have told us how they've been encouraged to learn how many of their strengths are connected to their dyslexic cognitive style.

The second thing that can help dyslexic individuals form a positive self-concept is to learn to regard your strengths as your true defining features. While it's important to identify your weaknesses so you can deal with them, it is even more important to identify what you do well. This includes any of the dyslexia-associated MIND strengths we've mentioned in this book, as well as any other important strengths you have, including interpersonal or leadership skills, artistic or athletic abilities, mathematical talents, or skills in any other significant area. The point is to put these strengths front and center in your model of who you are.

The third key to forming a positive self-concept is to accept dyslexia's dual nature. As we've shown, dyslexic differences arise from a pattern of brain organization that carries advantages in some areas and challenges in others. Don't shy away from admitting either to yourself or others any ongoing areas of challenge. Use strategy, technology, and collaboration to make up for those challenges, but don't let them chip away at your sense of self. Recognize that your challenges are part of a package deal. They are the price you pay for your strengths in other areas. Learn to focus on those strengths as the core of your identity and to deal with the challenges as the need arises.

Maintain a Positive Attitude

Attitude also has a major impact on self-concept. Staying positive in your outlook and beliefs about your future is both essential and entirely justified for individuals with dyslexia. Dyslexic minds are designed to thrive in adulthood. Two of the cognitive skills that are most useful for dyslexic thinkers are late-blooming in their development: the ability to use conscious working memory to make up for weaknesses in automatic processing, and the ability to use experience-based problem-solving networks. These abilities typically don't reach full function until the mid- or late twenties—for some people even the thirties. Be patient. Stay positive. Never give up. Time is on your side.

The greatest obstacle stopping dyslexic individuals from reaching their full potential is the loss of belief in a positive future. Too many dyslexic students, worn down by misunderstandings and the negative messages they have received, give up on themselves before the combination of developmental maturation, experience, and training have reached the phase of swift maturation that usually occurs somewhere between early adolescence and young adulthood. This loss of hope is a far greater impediment than any difficulties with reading or writing. Unlike the reading and writing challenges that accompany dyslexia, which can be managed with appropriate training, technology, and tactics, this loss of belief cannot be compensated for by any other means.

Fortunately, there is always reason for hope. Once the brain systems that support mature cognitive function have reached the necessary level of development, then the seeds of experience and training that have been planted and watered for so long will finally produce shoots.

In addition to hope, another thing is needed to keep you going during these tough times: the kind of mental toughness that goes by names like resilience, grit, or perseverance. This toughness allows you to tolerate short-term frustrations and failures without abandoning your long-term commitment and effort. It helps you decide that no matter how many times you stumble or fall, you will get back up and try again, and you will never stop trying until you succeed. It helps you see that what is important for your ultimate success is not the immediate outcome of your current efforts, but the continued and diligent pursuit of your long-term goals.

Although some individuals are born with more resilience and grit than others, these are traits that everyone can cultivate—and individuals with dyslexia are usually provided with plenty of opportunities to do so. Many successful dyslexic adults attribute their abilities to withstand setbacks in their adult professional lives—in business or sales or science or politics or sports or acting or countless other professions—to their early and repeated experiences of dealing with frustration and failure that they experienced growing up dyslexic. As we have also mentioned, the dyslexic bias toward error-driven learning helps dyslexic individuals recognize that errors themselves are not something to be avoided at all costs, but something that should be embraced, attended to, and learned from. Embracing setbacks both for the lessons they provide and for the opportunities to develop the key personal habits of resilience and grit is extremely important for all individuals with dyslexia. Parents and teachers should also take care to cultivate the development of these attitudes in the young dyslexic people in their care.

Select and Shape Your Supportive Environment

Your self-concept can also be affected to a great degree by your environment—primarily by the feedback you receive from those around you and by the success you personally experience. That's why it's so important that you take an active role in choosing and shaping your environment. Even well-intentioned teachers or employers aren't likely to have a deep understanding of how to accommodate dyslexic challenges or how to nurture dyslexic strengths. Dyslexic individuals and the parents of younger dyslexic students must become knowledgeable, outspoken, and persistent advocates, so they can supply this missing information and help shape an educational or workplace environment that will allow them to flourish and to exercise their strengths.

The goal of such advocacy should not be simply to get *out* of unproductive tasks, though often that is part of the conversation. Instead, it should be getting *into* areas that are well suited for your personal strengths. This is where a clear-eyed understanding of your abilities and needs will be so important, as will an understanding of the needs and goals of your employers, coworkers, and educators. Fortunately, creating complex mental models that can take such competing needs and goals into account is something dyslexic minds usually do well, and it can be the key to creating win-win situations.

Next, learn to understand and play your own game. Discover the roles, tasks, and situations that bring out the best in you—the settings where your strengths can play a valuable role and your interests and motivations are aligned with the work. Stick to activities that favor your own advantages. Many problems at work arise because of overly rigid requirements regarding means and methods, rather than because of difficulties meeting the desired ends. Choose workplaces that offer ends-focused policies and flexibility regarding means.

Be on the lookout for settings and people that seem to appreciate and understand your strengths, rather than judging you by your weaknesses. Because the best opportunities for flourishing as a dyslexic person will often be found in collaboration with others, finding partners, teammates, employers or employees, coaches, instructors, and others who can truly *see* you—and minimizing your contact with those who can't—will be important in finding success and maintaining your healthy self-concept.

Well Begun Is Half Done: The Importance of Family

For individuals with dyslexia, the process of building a strong self-image should ideally begin at home. One of the most wonderful things we've seen in our work with dyslexic families is the profoundly healthy impact that parents who have a positive vision of dyslexia can have on the developing self-concept of a dyslexic child. This doesn't remove all the frustrations or difficulties that accompany dyslexic challenges—especially in the early years—but it does create a larger and more favorable context for viewing these challenges and for creating resilience and a belief in a bright future.

In the rest of this chapter, we'd like to introduce you to a mother and daughter who've been through this journey together and show you how the fruits of their experience in creating a positive view of dyslexia are now being passed on to a whole new generation of dyslexic children. We first met Valerie Echavarria shortly after the first edition of this book was published, when she contacted us through our Dyslexicadvantage.org nonprofit. We were deeply impressed by her energy and enthusiasm, and by her commitment to helping not only her own children but also the broader dyslexic community. In the paragraphs that follow, we'll share with you first Valerie's perspective as a mother of dyslexic children, followed by the perspective of her daughter, Amparo.

Valerie's View

"I remember the first time we took Amparo for an evaluation by doctors. They acted like they were giving me a death sentence—almost like they were telling me she had leukemia. They told me that she could never learn to read, so I asked the doctor, 'Not even when she's in high school?' And the doctor told me no.

"But I went on the Internet and I began to read, and I learned that there were many things that I could do to help my child with dyslexia. Eventually I also found books like *The Dyslexic Advantage,* and as I read more, I thought, *Oh my, that's my dad. He thinks in just that way.* He was a school dropout who did not know how to read or write, but he was also an inventor who invented three manufacturing machines and had a company in Colombia that employed seven thousand workers. And when the president of Colombia announced he would give him a special prize, my father could not even write down his own speech, so he dictated it to a friend. We have many other people in our family who are builders and architects as well, and some of them are dyslexic, too—like my husband's brother, who had failing grades through nine years of school, but eventually became the top student in his design program at university and is now one of the most successful architects in Colombia. My husband is also dyslexic, and I am dyslexic and have ADHD. So it is probably no surprise that all four of our children turned out to be dyslexic. But my research helped me to realize that I am not a weirdo, which is what I had always thought when I struggled so badly in school. I am part of a special society—of something important. I would always tell my children that although they might hear bad things about dyslexia, or be told that they must be stupid and lazy if they can't easily read, just look at your family members and you will see that it is possible to shine with dyslexia!

"Over time I became involved in the dyslexia world and attended conferences like those of the International Dyslexia Association, where I learned so much. I learned about Orton-Gillingham intervention, and Structured Literacy, and how important systematic instruction is for learning. I also made wonderful contacts. One person I learned a lot from was Bill Keeney, who stressed the importance of 'ear learning,' or using recorded or text-to-speech books for students who still can't read well enough to learn all their lessons by reading themselves. He told me about resources like Learning Ally and Bookshare.org that provide recorded and text-to-speech books, and that was just crucial for my kids, because I couldn't always read well enough myself to read all their lessons for them until they could read well enough on their own.

"Finding that balanced approach between learning to read and using assistive technology to fill in the gaps and keep learning alive was so important for my kids. And I really had to be their advocate, because this approach was new to them. I would go to the parent-teacher meetings at school and they would say to me, 'You know, Amparo is still reading below the grade level,' and I would say, 'Yes, we are working on that, but please let her use her recorded books, because what I care most about for now is her love of books and her love for learning, and she learns well by listening.' So that assistive technology was essential in helping her to learn and do well while she was still improving her reading. Using that balanced approach, by middle and high school she was able to make the honor roll every term.

"It was not always easy, of course. All of my kids cried a lot when they were little. They would say, 'Mom, why do I study so hard and get a 70 or 60, and my classmate doesn't study at all and gets 100?' And I would tell them, 'Someday you're going to get those numbers, too, but for now some of the most important things you are learning are perseverance and resilience—you are learning how to fight for things. It is so easy now for your friends, but when they get to high school and they face something difficult and start making mistakes, then they're going to cry because they haven't learned to fight like you have, so don't worry.' So they were very resilient, and they also became very empathetic and helpful for other people who struggled."

Amparo's View

"To be honest, I really didn't love being dyslexic at first. My mother made me work very hard and it was often very difficult and frustrating. The challenges weren't just with reading. I remember that vocabulary tests were very hard, and I really hated them. I did so bad—I couldn't remember anything. I would study for hours, and that was such a big struggle for me. I felt like I couldn't do a lot of things.

"But my mother would also often emphasize that dyslexia was a gift, and she made me feel like it was something I shouldn't be ashamed of. Sometimes as a young student I would go to school and tell the other kids, 'Yeah, I'm dyslexic,' and then some of the kids would come to me quietly and say, 'I'm also dyslexic, but why are you saying it out loud?'

"For me, I know that it was having parents who understood dyslexia that helped me so much. Their attitude wasn't: 'Oh, man, you're not reading, and this is so horrible.' Instead, they said, 'Okay, you're not able to read this assignment, so let's try and figure out other ways that you can learn it.' And that kind of support and flexibility was so important for me, but it was just not something I received in school. It was still hard being a student with dyslexia, but just knowing that my parents were there to support me and were always trying to understand my thought processes really helped. They were also trying to understand their own thought processes, because both of them are dyslexic, and I think it was a really neat process of self-discovery for them, also. So together we really formed a partnership.

"Despite all the support I got, learning to read was still very hard. I had a lot of different types of interventions, but at first I didn't make much progress. I didn't ever have summer vacations: I just had intensive reading intervention. It wasn't until the summer when I was going from third grade to fourth grade that I finally had a teacher who changed my life, and it wasn't so much the instructive technique that she used that made the difference, but really just this way she had of making me believe that I could do it. I would go to her house and basically just play with her for a while before starting on reading. And I remember, for example, that she had me play the piano, and I didn't really know how to play it well, but she told me that I was doing a great job learning how to play it, and somehow that made me feel like a learner—like I was achieving something in the learning environment, which was something I hadn't ever felt before. And that feeling of being a learner carried over when we started working on reading. And although I certainly wasn't reading fluently at the end of summer, when I went back to school in fourth grade, I wasn't afraid to grab books and try to read a page or two. So that's really when I started to read a bit more.

"But for a long time I still didn't learn that well through reading, and to be honest, it's still not my favorite way to learn, even though that's the way most schools want you to learn: just read the passage and learn the material. I learn better by listening than reading, and I learn best of all by doing things and learning from experience. It actually took me quite a while to realize that about myself, because I had absorbed this idea at school that I really needed to learn everything by first learning how to read well enough so I could understand what I'm reading. And it really wasn't until college that I realized clearly that the way I learn best is by practical experience and doing things firsthand.

"When I started college, I asked myself, *What am I really good at?* And I realized that in high school the thing I was best at—and what I also enjoyed most—was my community service project. In middle school and high school my brother and I created a group called Feeding Souls, where we would organize volunteers, donations, health services, and the opportunity for rehabilitation for people experiencing homelessness, drug addiction, or alcoholism in the neighborhood surrounding our school. I also developed a mentoring and self-advocacy program by and for students with dyslexia. I learned so much from those experiences just by having conversations with people, and those were really the best learning experiences I ever had.

"So when I started college, I thought maybe what I should do was social work, because that seemed most like those experiences. But after my first year of college, I realized that my favorite class was educational justice, so I thought, 'Okay, even though school was so hard for me, I'm going to be a teacher, and just hope for the best.'

"It actually turned out perfectly for me because just by coincidence the school I went to, Loyola University Chicago, taught through experience-based learning, so all of my classes were experience-based. My very first class involved going to a school and observing the classroom, and I was so thrilled because I was like, 'I'm learning, you know, I'm actually learning!' And when I did my student teaching, I was learning all these things, and I got certifications in special ed and in Wilson intervention, and every day I was just learning and learning. Before college, that just never happened in my classes, to be honest.

"I also really loved the school where I did my student

teaching, and I actually went to work there for a year after graduation. I worked there with kids with dyslexia, and I got to help them find and develop their own strengths. I also got to help them use tools for their challenges, like assistive technology, and to learn self-advocacy. I really enjoyed that, and the effect on the kids was magical. They would come from schools where they had encountered really bad circumstances and misunderstanding, so the first day they would come into our school they would be like, 'I hate school, it's horrible.' But they soon found out that they were now in a safe and encouraging place. I loved sharing with them that I also was dyslexic, and proud of it! I told them that I also had the same problems they did, but that I had strengths as well, and so did they. And in the end they were like, 'I love learning, and I love my brain!' which was so exciting!

"Like my mother said, the need to work as a team between parents and teachers and students is just so important, and I see the importance of this all the time now as a teacher. Some parents just want to hand over their child to the school or a tutor and say, 'You fix the problem.' But other parents say, 'Okay, what do *we* have to do? How can we help you? What plans are we going to create to help our child succeed? How can we use their talents?' That makes such a huge difference! For kids where support at home is lacking, those kids are usually not emotionally ready for the struggle. There's no motivation, because while they may get support at school, when they get home to do their homework, no one is there to support or encourage them or to point out what they are good at. The parents may look at the grades and the test results, but they're not a part of the learning process. But it really does take a team.

"I also think that for many people there's still a very big misunderstanding of what dyslexia is. It's not just about reading intervention, and that's it. It's more than just that. It's writing and memorization and other challenges, too; but there are also the talents, and those are too often being wasted.

"Unfortunately, I think it's especially sad when I talk to Latin American parents, because they almost always think of dyslexia as a cognitive disability, and they fear that being diagnosed will harm or impair their kids. As a result, they don't want their kids to know that they're dyslexic, and they don't want to label them in that way. It is really a big stigma, much bigger than in the United States.

"In fact, I did a little conference in Colombia recently, and it was so interesting to see the people's reactions when I talked about my dyslexia—I was actually surprised at the questions they asked. They almost all assumed that dyslexia meant you could never learn to read or graduate from high school, and they just couldn't understand how I could have dyslexia and still have been able to graduate college and become a teacher. Some of them also had a hard time understanding how I could talk about it openly and even be proud of it, because the awareness that dyslexic people usually have many strengths is just not there.

"Even for Spanish-speaking children in the United States, the assumptions parents and sometimes teachers make can often be very limiting. They will often assume that if it is already hard for a child to learn to read in Spanish and if English is even harder, then bilingual education will just be impossible. So they sometimes choose to avoid English altogether, but this can be very limiting later in life. Often the bilingual child's dyslexia is missed entirely, because the assumption is that any reading problems are simply caused by the bilingualism. But both of those ideas are just not right.

"In the end my goal is to take the things I have learned about dyslexia here in the U.S. back to Latin America and to help young people with dyslexia there, because the need is so great and there is so much misunderstanding. The idea that dyslexia is a disadvantage has been holding back so much talent and growth from happening in the Latin American countries. My goal is to help my community see the advantage that dyslexia brings, and to stir a change in families, schools, and the Latin American community as a whole. A new mindset is the true gateway to every dyslexic person's success, and a healthy self-image and the understanding of dyslexia as a gift is the key to the dyslexic advantage."

Epilogue

We are often asked how we first decided to write a book about dyslexic strengths. There were two main reasons. The first was the overwhelming evidence of these strengths that we saw in the dyslexic people with whom we worked. In our learning clinic, we both participated in testing every child and adult we saw. When one of us was in the exam room, the other was outside talking with the family. As a result, we ended up getting unusually detailed histories not only about the examinee—their challenges, strengths, interests, goals, attitudes, feelings, developmental patterns, etc.—but also about the extended family. Over time, it became clear that the same strengths and challenges we were seeing in the exam room were closely mirrored by the real-world strengths and challenges shown by many family members. It was this dual heritability—both strengths and challenges—that first persuaded us that the deficit model of dyslexia was inadequate to explain the full range of dyslexia-related cognitive features we were witnessing. We came to believe that dyslexic minds were better understood as optimized for a different set of functions, and that this optimization created both challenges and strengths. Our convictions were strengthened by further analyses of the scientific literature both on dyslexia and cognition in general and by our studies of the lives of many dyslexic adults. We found this information so exciting and relevant for the lives of dyslexic people—and the non-dyslexic world as well—that we wanted to share it with others.

The second reason was very different, and even more urgent. It was the damage we saw being done to so many wonderfully talented and curious dyslexic children as they entered into educational systems that were ill equipped to nurture or even understand their unique developmental needs. By nature, these children were deeply curious, and usually began school full of anticipation, eager to learn and explore. However, when faced with a system that had not been designed with their learning needs in mind, they soon began to feel defective and broken. Before long, many became anxious, withdrawn, or agitated, and filled with self-doubt. We realized that this suffering was as unjust as it was unnecessary—that it resulted simply from a failure to understand and provide for an entirely healthy cognitive pattern. We were determined to do what we could to address this injustice.

In the years since *The Dyslexic Advantage* was first published, the case for dyslexic strengths has only grown stronger and the nature of those strengths clearer. Although our educational systems have barely begun to implement the changes that are needed to support dyslexic learners, an increasing number of schools and nonprofit organizations around the world have begun to advocate for a more strengths-based approach to dyslexic learning and employment, including our own nonprofit organization, Dyslexic Advantage (dyslexicadvantage.org), which has grown to over 100,000 members. We have been especially thrilled to observe how the involvement of loud and proud dyslexic individuals has played an increasingly valuable role in this process—including the people who have been profiled in this book.

We are also beginning to see increased interest from many employers, who want to learn how they can benefit from the tremendous resources of creativity and talent that dyslexic individuals possess. We are finally reaching the point where we can begin to offer explicit strengths-based best practices for workplaces, as well as schools and training programs. Developing such guidance should be one of the highest priorities of the dyslexia movement going forward, and in doing so we should follow the advice of the visionary—and almost certainly dyslexic—architect and urban planner Daniel Burnham, whose favorite saying was "Make no small plans."

The first step in planning for a world that will be better for dyslexic individuals is to completely change our thinking about what it means to be dyslexic. We must bring about the "Copernican revolution" mentioned early in this book, removing dyslexia-associated challenges from the center of our model of dyslexic cognition and replacing them with dyslexia-associated strengths—just as Copernicus replaced the earth with the sun at the center of his heliocentric model. These dyslexia-associated strengths—and the advantages provided by them to the human community as a whole—are both the reason why so many of us are dyslexic and the key to understanding how dyslexic minds learn and work to their fullest potential.

Most scientific revolutions don't require irrefutable proof, but simply a recognition that the new view better explains all the relevant data than the old one. Copernicus never proved experimentally that the planets revolved around the sun; he simply showed that a sun-centered model more simply and adequately explained the observable motions of the planets than the old earth-centered one. Therefore, the question we should ask is whether a strengths- or a deficit-centered model of dyslexia better explains why so many struggling students can go on to succeed as adults; why a student like our friend David McComas can fail to learn to read until age twelve, then become a professor of astrophysics at Princeton and lead billion-dollar NASA missions; why college dropouts like Jack Horner can revolutionize paleontology and win genius awards; and why special ed students like Carol Greider can win the Nobel Prize in Medicine.

There is clearly a radical incoherence at the heart of the deficit-centered model of dyslexia because it fails to explain what we can see with our own eyes—if we would only look. But once you view dyslexia from a strengths-centered perspective, you can understand both the greater power of this new perspective and the limitations of the old one.

As hopeful as we are that this reappraisal of dyslexic minds will continue to gain force, this change will not take place overnight, and tomorrow millions of dyslexic individuals must return to spend time in classrooms and workplaces where this understanding has not yet taken hold. In emphasizing dyslexic strengths, we have no wish to downplay the very real challenges that individuals with dyslexia face. But as we said at the beginning of this book, "suffering from dyslexia" is suffering of a very special kind. Rather than the suffering of a person with an incurable disease, it is the suffering of a hero on a perilous but promising quest.

As we noted in our first edition, perhaps the hero who best represents this dyslexic quest is Aragorn, from J. R. R. Tolkien's *The Lord of the Rings.* Early in the Ring saga, Aragorn seems little more than a wandering vagrant. Yet his fate is foretold in the lines of an ancient prophecy, which reminds us that things are not all as they first appear, and that royal natures sometimes lie hidden beneath rags:

All that is gold does not glitter,
Not all those who wander are lost

Aragorn is in fact the rightful heir to the throne of Gondor. Though crownless now he is destined one day to be king, when the broken sword that is his birthright shall be forged anew so that he can carry it into battle. If you are an individual with dyslexia, this is your story as well. Its truth does not rest on the words of an ancient oracle but on up-to-date scientific research, firsthand observations, and the experiences of the countless talented and successful individuals with dyslexia. Although you may not glitter in the same obvious ways that nondyslexic individuals do, you have a special brilliance all your own. If you devote the effort necessary to understand and develop your strengths, to manage your challenges, and to retain a positive view of your future potential, your dyslexic mind will be forged as keen as any blade. In the end, you'll find that rather than being "cured" of dyslexia, you'll become a perfect example of what an individual with dyslexia was always meant to be. And when you do, you'll understand the truth, and the true nature, of the dyslexic advantage.

Acknowledgments

This book would not have been possible without the help of the many wonderful people whom we now take great pleasure in thanking.

Our first and greatest thanks go to the thousands of dyslexic individuals and their families with whom we've been privileged to work over the last several decades. This book consists of lessons you've taught us, and nothing gives us greater pleasure than to hear from you that our work has made a difference.

To our agent, Carol Mann, who for more than fifteen years has been our constant advocate, occasional psychotherapist, and wise adviser: Thank you so much for all you do.

Thanks also to the amazing team at Dutton/Plume who have overseen this update and guided us through each stage of the revision: Ben Lee, associate publisher of paperback and backlist publishing, and Aurora Slothus, assistant director of backlist publishing; Jill Schwartzman, editorial director for Plume; Marie Finamore, senior production editor; Jamie Knapp, Plume publicity director; Jason Booher, art director, Tiffany Estreicher, design director; Jillian Fata, subsidiary rights manager; Nancy Inglis, copyeditor; and finally to editorial assistant Grace Layer, who has added immeasurably to the strengths of this book through her oversight and editing of our work.

Our profound thanks as well go to all who have supported and participated in our work with the Dyslexic Advantage nonprofit. First, to the members of the Dyslexic Advantage board, who have generously donated their time, talents, and insights toward helping this organization fulfill its mission of promoting the positive identity, community, and achievement of dyslexic people by focusing on their strengths: Thomas G. West, a true pioneer in this field, who has served on the board faithfully from day one, Joan Bisagno, Erin Egan, David Flink, Cinthia Haan, Kristi Helgeson, Kurt Heusner, Claudia Koochek, David McComas, Gerald Rittenberg, Nathan Stooke, Tanya Wojtowych, and Linda Yates. Our deepest thanks also to the countless donors whose financial support has enabled our organization to work toward its vision of creating a world where dyslexic people are known for their strengths. Special thanks to Kurt and Tammy Mobley, Gerry Rittenberg, Liz Steinglass, the Rauch Foundation, the Sang family, and Scott Sandell. Our great thanks also to the dedicated group of volunteers who have assisted Fernette with the newsletter, premium magazine, and other nonprofit activities: Nicole Swedberg, Trish Seres, Shelley Wear, Michelle Williams, Dayna Russell Freudenthal, Jennifer Plosz, Cheryl Kahn, Jack Martin, and Marcia Ciorra.

We would also like to thank Christina Schneider for her remarkably generous donation of her time and expertise in performing all the complex statistical work for the MIND strengths research presented throughout the book. This work could simply not have been done without her, and we are immensely grateful for her help.

Finally, there would be no second edition of this book were it not for some very special people who helped us through the loss of our daughter, Karina, to cancer in 2014. These people include our parents, Harry and Marian Fang, and Leonard and Ruth Marie Eide; our son, Krister; our dear friends and first responders Eryn Lee and Eric Aguiar, who ran toward our fire; Stephanie, Richard, Alan, and Rosalyn Bannon, whose kindness and hospitality we will never forget; Gerry and Linda Rittenberg, whose special act of kindness came at just the right time, and helped more than they will ever know; Laura Shawver, who gave many hours and her tremendous professional expertise to help identify possible therapies for Karina; Monica Banks, whose partnership with Fernette in the Karina Eide Young Writers Awards has been a profound source of joy; and Kaja Winiarz, who helped Brock to write again. To each of you we can never give thanks enough, but you will share in the thanks of every person who benefits from this new edition, for without you it would not exist.

Appendix: MIND Strengths Survey Results

For our MIND strengths surveys, we initially emailed an invitation and Internet link to 12,291 members of the Dyslexic Advantage Forum, which consists primarily of dyslexic individuals and their family members, teachers, tutors, and assessment professionals. Some 2,293 adults over the age of eighteen completed at least one of the four surveys, one for each of the MIND strengths. The surveys contained 91 questions in all. In total, 5,070 surveys were completed. Participants were asked to assign themselves to one of four dyslexia risk levels; answers from two of these groups are presented here: Definitely Dyslexic (i.e., had been formally identified as dyslexic on the basis of a professionally administered assessment), and Definitely Not Dyslexic (or showing none of a list of common dyslexia-associated symptoms). After analysis, 64 questions total were selected as showing both significant differences in the responses between dyslexic and nondyslexic participants, and high correlation to the respective MIND strength concept by principal component analysis.

In the tables following, we list the results for each of the four

MIND strengths surveys in separate tables. In each table we list the questions in the leftmost column. Then in the right four columns we list "Strong Agreement" (SA) and the "Strong Agreement + Agreement" (SA+A) first for the dyslexic and then for the nondyslexic respondents.

M-Strengths

Statement	Dyslexic Strongly Agree	Dyslexic Strongly Agree + Agree	Not Dyslexic Strongly Agree	Not Dyslexic Strongly Agree + Agree
1. I am very good at forming 3-D spatial images in my mind.	53	79	9	25
2. When I form 3-D spatial images in my mind, I can manipulate them at will and view them from all angles.	44	76	6	17
3. I have very good 3-D spatial reasoning ability	43	74	7	23
4. I can create an image in my mind of the working parts of machines and how they operate.	33	54	4	16
5. I have always been good at building things. [For example, as a child: toys like LEGOs, robotics kits, trains, K'NEX, marble runs, models, 3-D arts and crafts projects; as an adult: things like landscaping, home repair or renovation, machine or engine work, furniture building, sculpture, etc.]	52	74	7	24
6. When assembling a kit (e.g., furniture, model), I usually don't have to read the written instructions. I can just tell how things must go together by looking at the pictures.	43	73	7	21

M-Strengths

Statement	Dyslexic Strongly Agree	Dyslexic Strongly Agree + Agree	Not Dyslexic Strongly Agree	Not Dyslexic Strongly Agree + Agree
7. I generally prefer diagrams or pictures to written instructions or explanations.	70	90	15	42
8. I am very good at reading blueprints and at imagining in my mind the 3-D structures they represent.	39	68	5	18
9. When I think through a problem my thinking is more nonverbal (visual images, other sensory images, movements, etc.) than verbal (words).	53	84	8	21
10. Before I can describe what I think about something, I often have to translate my thoughts into words (that is, up to that point, my reasoning process hasn't primarily used words).	44	72	4	20
11. I'm especially good at reading topographical maps and can easily imagine the landscapes they represent in 3-D.	31	53	4	19
12. After going someplace once, I usually don't need directions or a map again to find it or to find my way home.	36	64	14	35
13. I can picture in my mind the 3-D layout of places where I've been and walk through them in my mind.	48	78	12	34

I-Strengths

Statement	Dyslexic Strongly Agree	Dyslexic Strongly Agree + Agree	Not Dyslexic Strongly Agree	Not Dyslexic Strongly Agree + Agree
1. I often see connections or relationships that other people miss (e.g., how things, people, events, etc., resemble each other, are alike, or are in other ways related or connected).	68	93	26	65
2. I often spot things or ideas that are missing or lacking (negative space thinking).	50	85	16	47
3. I am better at understanding the big picture than at thinking about details.	58	81	19	40
4. I often think of analogies and metaphors to better understand ideas or to innovate.	56	84	23	56
5. I tend to see relationships as complex webs rather than as links or chains.	55	82	25	56
6. I am good at detecting connections or relationships that make up a larger system.	55	89	20	61
7. I am good at detecting patterns in complex events or data sets.	39	71	14	38
8. I am good at approaching problems from different perspectives by switching my assumptions.	55	85	20	52
9. I am good at seeing things from another person's point of view.	51	82	36	77
10. I especially enjoy big-picture topics that try to explain the patterns behind what happens in the world, like science, philosophy, history, political science, theology, sociology, psychology, economics, etc.	67	88	30	66

I-Strengths

Statement	Dyslexic Strongly Agree	Dyslexic Strongly Agree + Agree	Not Dyslexic Strongly Agree	Not Dyslexic Strongly Agree + Agree
11. I am especially good at spotting the gist or the heart of the matter in an argument, story, or set of events.	60	89	29	74
12. When I am taught a new task, I can't just learn it by rote, but have to make sense of it before I can learn it.	72	94	27	60
13. I learn best if I start with the big picture or general overview before trying to master the details.	68	93	30	66
14. When I study a new subject, I'm often confused about the importance of the details until it suddenly all clicks and I can see the big picture.	56	86	14	41
15. I often have difficulty learning something until I can understand its point, or why I am being asked to learn it.	58	89	19	49
16. I am good at coming up with new uses for things (that is, using them in ways for which they weren't originally intended).	48	83	10	34
17. I often combine approaches or techniques from different disciplines.	54	88	22	62
18. In my work I often use approaches or techniques that I came up with myself.	57	87	19	59
19. I often use techniques or approaches for applications that are different from those for which I was taught to use them.	46	82	14	45

N-Strengths

Statement	Dyslexic Strongly Agree	Dyslexic Strongly Agree + Agree	Not Dyslexic Strongly Agree	Not Dyslexic Strongly Agree + Agree
1. When I try to recall a fact or a concept, the first thing that comes to my mind is usually a case, example, personal experience, or image, rather than an abstract verbal definition.	68	93	30	65
2. When I'm trying to learn a new concept, I prefer my instructor to start with a story, case, or example, rather than to start with a precise definition.	58	87	31	64
3. I typically reason, remember, and learn better using stories, cases, or examples, rather than abstract concepts or definitions.	61	88	33	70
4. When I recall my past, I don't just remember naked facts about who, what, when, where, why, and so forth, but I reconstruct my old experiences in my mind in such vivid detail that it's almost like I'm experiencing them again.	59	84	24	50
5. I often need to use special tricks to memorize rote information (naked facts or lists), like making up a story, song, or rhyme about them.	46	71	16	39
6. When I think of historical facts or events, I see scenes or depictions in my mind, rather than just recalling verbal descriptions.	54	74	13	47

Appendix: MIND Strengths Survey Results

N-Strengths

Statement	Dyslexic Strongly Agree	Dyslexic Strongly Agree + Agree	Not Dyslexic Strongly Agree	Not Dyslexic Strongly Agree + Agree
7. If someone asked me to think about an abstract term like *justice*, I'd usually think first about stories or images or examples of just behavior, rather than abstract definitions of justice, like "giving everyone their due" or "treating everyone the same."	51	82	14	43
8. If someone asked me to define *tree* or *car*, the first thing my mind would do is imagine a whole series of trees or cars, rather than jump directly to an abstract, word-based definition.	61	85	19	47
9. I usually remember things I'm told in one-on-one conversations much better than things I hear in lectures or read in books.	41	78	10	38
10. I'm much better at remembering things I've done than things I've simply been told about.	68	92	29	70
11. I think I notice and remember more things that I encounter by chance in my environment than most other people I know.	49	76	15	36
12. I often seem to remember things I encounter by chance experience better than the things I tried hard to memorize in school.	55	84	13	42
13. I learn much better by practical experience than by formal instruction (e.g., books, lectures, paperwork).	68	90	19	50

D-Strengths

Statement	Dyslexic Strongly Agree	Dyslexic Strongly Agree + Agree	Not Dyslexic Strongly Agree	Not Dyslexic Strongly Agree + Agree
1. I am good at predicting how current events or processes will unfold and what is likely to happen next.	39	77	10	42
2. It often seems obvious to me what direction events will take in the future.	36	75	10	42
3. I often see several steps ahead of those around me.	49	81	15	43
4. When I watch a show or read a story, I can often predict what will happen next or how the plot will develop.	49	80	28	66
5. I often spot new and developing trends before others.	34	70	8	34
6. I am good at connecting the dots and spotting patterns or processes that others miss.	56	86	17	52
7. I am good at mentally simulating (or imagining) how a complex system or process will work, and how changes in conditions and components will affect their outcomes.	46	77	7	29
8. I am good at troubleshooting complex systems or processes and finding out why they're not working or where things went wrong.	36	70	9	33
9. I am good at developing creative strategies for reaching difficult goals or solving complex problems.	51	83	14	40
10. I often seem more comfortable than most other people in situations where there are a lot of unknowns or where things are changing rapidly.	41	70	13	38

D-Strengths

Statement	Dyslexic Strongly Agree	Dyslexic Strongly Agree + Agree	Not Dyslexic Strongly Agree	Not Dyslexic Strongly Agree + Agree
11. I often seem to have a better sense of what to do than most other people in situations where there are a lot of unknowns or where things are changing rapidly.	40	77	11	38
12. I'm at my best in situations where there are lots of unknowns or where things are changing rapidly.	27	58	4	21
13. My mind often solves problems or reaches the answers to complex questions through sudden insights, intuitions, or hunches.	60	89	18	51
14. When I solve math problems I often see the solutions suddenly, then have to go back to figure out the steps to reach them.	16	36	2	7
15. To answer a difficult problem, I often have to switch gears and take a break or work on something unrelated.	33	71	10	40
16. I spend a lot of time daydreaming and letting my mind wander.	46	76	9	29
17. Some of my best work is done when I'm daydreaming or letting my mind wander.	43	74	8	28
18. Other people often find my thinking nonlinear.	57	85	10	27
19. I often have difficulty explaining to other people how I come up with the answers I do.	51	77	5	19

Notes

Chapter 1. A New View of Dyslexia

8 **as many as one in five individuals:** Current research suggests that as many as 20 percent of U.S. residents may be considered dyslexic. See, e.g., https://dyslexia.yale.edu/dyslexia/dyslexia-faq.

8 **"the smartest lad in the school":** W. P. Morgan, "A Case of Congenital Word Blindness." *British Medical Journal* 2 (1896): 1378; doi: 10.1136/bmj.2.1871.1378.

9 **"dyslexia is a specific learning disability":** https://dyslexiaida.org/definition-of-dyslexia/.

Chapter 4. Sources of Dyslexia's Dual Nature

34 **a steady stream of books and articles:** Widely read examples include Daniel Pink, *A Whole New Mind: Why Right-Brainers Will Rule the Future* (New York: Riverhead, 2005) and Betty Edwards, *Drawing on the Right Side of the Brain*, 4th definitive, expanded, updated ed. (New York: TarcherPerigee, 2012).

34 **process information in very different ways:** Iain McGilchrist, *The Matter with Things: Our Brains, Our Delusions and the Unmaking of the World* (London: Perspectiva Press, 2021).

35 **they specialize respectively in:** For those interested in a good general discussion of these differences, we recommend Robert Ornstein, *The Right Mind: Making Sense of the Hemispheres* (New York: Harcourt Brace, 1997), or McGilchrist, *The Matter with Things*.

35 **with dyslexia, the two brain hemispheres:** Albert M. Galaburda, Marjorie LeMay, Thomas L. Kemper, and Norman Geschwind, "Right-Left Asymmetries

in the Brain." *Science* 199, no. 4331 (1978): 852–56; doi: 10.1126/science.341314.

36 **author Thomas G. West:** Thomas G. West, *In the Mind's Eye: Creative Visual Thinkers, Gifted Dyslexics, and the Rise of Visual Technologies*, 2nd ed. (Amherst, NY: Prometheus Books, 2009).

36 **visual and spatial talents:** Dr. Maryanne Wolf, who is the director of the newly created Center for Dyslexia, Diverse Learners, and Social Justice at UCLA, also comments on this apparent dyslexia/right-hemisphere connection in her fascinating survey of reading and reading challenges, *Proust and the Squid: The Story and Science of the Reading Brain* (New York: Harper Perennial, 2007).

36 **This difference was first demonstrated:** S. E. Shaywitz et al., "Functional Disruption in the Organization of the Brain for Reading in Dyslexia." *Proceedings of the National Academy of Sciences of the USA* 95, no. 5 (1998): 2636–41. This pattern has also been confirmed by many other researchers. For a detailed discussion of this reading circuit, see Wolf, *Proust and the Squid*, 165–97, or Stanislas Dehaene, *Reading in the Brain: The Science and Evolution of a Human Invention* (New York: Viking, 2009), 235–61.

36 **Reading expert Dr. Maryanne Wolf:** Wolf, *Proust and the Squid*, 186.

36 **Dr. Guinevere Eden and her colleagues:** Peter E. Turkeltaub et al., "Development of Neural Mechanisms for Reading." *Nature Neuroscience* 6, no. 7 (2003): 767–73; doi: 10.1038/nn1065.

37 **the shift that takes place in expert musicians:** Ornstein, *The Right Mind*, 174.

38 **Dr. Mark Beeman published:** Mark Jung-Beeman, "Bilateral Brain Processes for Comprehending Natural Language." *Trends in Cognitive Sciences* 9, no. 11 (2005): 512–18; doi: 10.1016/j.tics.2005.09.009.

40 **"It produces these inferences by detecting":** The actual experiment by which Beeman demonstrated this is described in the *Trends* paper cited above and runs as follows. First, individuals were primed by being shown three different words (in this example *foot*, *glass*, and *pain*) that are each distantly related to a particular word (in this case *cut*). Second, the word *cut* was shown either to the left or the right hemisphere of the brain (by displaying it exclusively to either the right or the left half of the visual field). When this was done, only the right hemisphere responded more strongly than it did when the prime was not given, because only its broader semantic field created a cumulative priming effect through which the activation of each related word added additional force. In contrast, when a word more closely related to *cut*—like *scissors*—was used to prime subjects prior to viewing *cut*, it was now the left hemisphere that showed a greater priming effect. In short, the right hemisphere recognizes secondary or more distant semantic relationships that help capture overall meaning or gist, while the left hemisphere recognizes almost exclusively the "tight" primary meanings that help maintain precision.

41 **more heavily engage the right hemisphere:** Michele T. Diaz and Anna Eppes, "Factors Influencing Right Hemisphere Engagement During Metaphor

Comprehension." *Frontiers in Psychology* 9 (2018): 414; doi: 10.3389/fpsyg.2018.00414.

41 **In a fascinating study from Canada:** Pam Hruska et al., "Hemispheric Activation Differences in Novice and Expert Clinicians During Clinical Decision Making." *Advances in Health Sciences Education* 21, no. 5 (2016): 921–33; doi: 10.1007/s10459-015-9648-3.

42 **individuals with dyslexia and autism:** A useful summary of this work is provided in Emily L. Williams and Manuel Casanova, "Autism and Dyslexia: A Spectrum of Cognitive Styles as Defined by Minicolumnar Morphometry." *Medical Hypotheses* 74, no. 1 (2010): 59–62; doi: 10.1016/j.mehy.2009.08.003.

43 **Jeff Hawkins explains in his book:** Jeff C. Hawkins, *A Thousand Brains: A New Theory of Intelligence* (New York: Basic Books, 2021), 47, 50.

47 **"If you test patients with autism":** That is, autistic individuals tend to interpret messages based on a very narrow, literal, or "concrete" understanding of the words used, relying almost entirely on the primary word meanings.

48 **This correlation argues strongly:** Although many of the cognitive features associated with the bias toward long connections are similar to the features associated with increased right-brain processing that we described above, one advantage that Dr. Casanova's minicolumnar theory of dyslexia has over the right-brain-predominant theory that we discussed above is that it does a better job of explaining why individuals with dyslexia typically retain a right-brain flavor to their processing style even when brain scans show that their circuitry has become increasingly left-sided through practice. For example, we often find that individuals with dyslexia who've become relatively skilled readers still process stories in a highly gist-dependent, top-down fashion, just like many less-skilled dyslexic readers. We'll explain this finding in more detail in our section on I-strengths, but the basic point is that certain aspects of the dyslexic processing style are unlikely to completely vanish even with extensive training, as we might have predicted with the hemispheric theory, because the difference in minicolumnar orientation and bias toward long connections means that the *left* hemisphere of an individual with dyslexia will in some ways function with a rather *right*-hemispheric flavor.

Chapter 5. More Conscious, Less Automatic

49 **detailed model of dyslexic brain development:** Roderick I. Nicolson and Angela J. Fawcett, "Development of Dyslexia: The Delayed Neural Commitment Framework." *Frontiers in Behavioral Neuroscience* 13 (2019): 112; doi: 10.3389/fnbeh.2019.00112.

50 **At least half of all individuals with dyslexia:** Michael T. Ullman, F. Sayako Earle, Matthew Walenski, and Karolina Janacsek, "The Neurocognition of Developmental Disorders of Language." *Annual Review of Psychology* 71 (2020): 389–419; doi: 10.1146/annurev-psych-122216-011555.

50 **a large majority of dyslexic individuals:** See the results of the MIND strengths surveys presented in the "What's New" chapters in each of the four MIND strengths sections.

54 **dyslexic readers' eyes will commonly drift:** John Stein, "The Current Status of the Magnocellular Theory of Dyslexia." *Neuropsychologia* 130 (2019) 66–77; doi: 10.1016/j.neuropsychologia.2018.03.022.

54 **A similar process takes place with hearing:** Stein, "The Current Status of the Magnocellular Theory of Dyslexia."

56 **Rod and Angela have actually raised:** Nicholson and Fawcett, "Development of Dyslexia," 112.

Chapter 6. Dyslexia-Associated Challenges

61 **70 to 80 percent of individuals with dyslexia:** Gabrielle O'Brien and Jason D. Yeatman, "Bridging Sensory and Language Theories of Dyslexia: Toward a Multifactorial Model." *Developmental Science* 24, no. 3 (2020): e13039; doi 10.1111/desc.13039.

63 **60 percent showed problems with both:** Elizabeth S. Norton and Maryanne Wolf, "Rapid Automatized Naming (RAN) and Reading Fluency: Implications for Understanding and Treatment of Reading Disabilities." *Annual Review of Psychology* 63 (2012): 427–52; doi: 10.1146/annurev-psych-120710-100431.

63 **visual processing problems play a significant role:** O'Brien and Yeatman, "Bridging Sensory and Language Theories of Dyslexia."

69 **about 25 to 40 percent of dyslexic students:** Erik G. Willcutt and Bruce F. Pennington, "Comorbidity of Reading Disability and Attention-Deficit/Hyperactivity Disorder: Differences by Gender and Subtype." *Journal of Learning Disabilities* 33, no. 2 (2000): 179–91; doi: 10.1177/002221940003300206.

70 **"risk factors for dyslexia":** Primarily the six processing challenges we've just discussed.

70 **"wrote a paper a few years back":** Stephanie L. Haft, Chelsea A. Myers, and Fumiko Hoeft, "Socio-Emotional and Cognitive Resilience in Children with Reading Disabilities." *Current Opinions in Behavioral Science* 10 (2016): 133–41; doi: 10.1016/j.cobeha.2016.06.005.

72 **Although Mandarin symbols don't represent:** Chen Chen, Yue Yao, Zhengjun Wang, and Jingjing Zhao, "Visual Attention Span and Phonological Skills in Chinese Developmental Dyslexia." *Research in Developmental Disabilities* 116 (2021): 104015; doi: 10.1016/j.ridd.2021.104015.

Chapter 8. The Advantages of M-Strengths

82 **In the first study, British psychologist:** Elizabeth Ann Attree, Mark J. Turner, and Naina Cowell, "A Virtual Reality Test Identifies the Visuospatial Strengths of Adolescents with Dyslexia." *Cyberpsychology and Behavior* 12, no. 2 (2009): 163–68; doi: 10.1089/cpb.2008.0204.

84 **A second study, performed by:** Catya von Károlyi, "Visual-Spatial Strength in Dyslexia: Rapid Discrimination of Impossible Figures." *Journal of Learning Disabilities* 34, no. 4 (2001): 380–91; doi: 10.1177/002221940103400413.

85 **Dr. Jean Symmes, who was a research psychologist:** Jean S. Symmes, "Deficit Models, Spatial Visualization, and Reading Disability." *Bulletin of the Orton Society* 22 (1972): 54–68.

85 **Former Harvard neurologist Dr. Norman Geschwind:** Norman Geschwind, "Why Orton Was Right." *Annals of Dyslexia* 32 (1982): 13–30.

86 **A more direct link between dyslexia:** Psychologist Alexander Bannatyne noted that in his experience "parents in highly spatial occupations, such as surgeons, mechanics, dentists, architects, engineers and farmers, tend to have more dyslexic children than do those in other occupations." Alexander Bannatyne, *Language, Reading and Learning Disabilities: Psychology, Neuropsychology, Diagnosis and Remediation* (Springfield, IL: Charles C. Thomas, 1971).

86 **In the United Kingdom:** https://www.rca.ac.uk/more/about-rca/official-information/college-policies-and-codes-of-practice/dyslexia-policy-new/.

86 **At the Central Saint Martins College:** Beverley Steffert, "Visual Spatial Ability and Dyslexia." In *Visual Spatial Ability and Dyslexia*, ed. Ian Padgett (London: Central Saint Martins College of Art and Design, 1999), 10–49.

86 **Harper Adams University in England:** https://www.lboro.ac.uk/media/media/schoolanddepartments/mathematics-education-centre/downloads/events/2004-eng-maths-dyslexic-student/sarah_parsons_notes.doc.

86 **In Sweden, researchers:** Ulrika Wolff and Ingvar Lundberg, "The Prevalence of Dyslexia Among Art Students." *Dyslexia* 8, no. 1 (2002): 34–42; doi: 10.1002/dys.211.

86 **In his book *Thinking Like Einstein*:** Thomas G. West, *Thinking Like Einstein: Returning to Our Visual Roots with the Emerging Revolution in Computer Information Visualization* (Amherst, MA: Prometheus Books, 2004).

87 **UCLA professor Maryanne Wolf:** Wolf, *Proust and the Squid.*

87 **And author Lesley Jackson:** https://www.iconeye.com/icon-013-june-2004/dyslexia-icon-013-june-2004.

87 **Dr. Norman Geschwind wrote:** Geschwind, "Why Orton Was Right."

87 **neurologist Macdonald Critchley:** Macdonald Critchley and Eileen A. Critchley, *Dyslexia Defined* (London: W. Heinemann Medical Books, 1978).

88 **The Mosers identified cells:** May-Britt Moser, David C. Rowland, and Edvard I. Moser, "Place Cells, Grid Cells, and Memory." *Cold Spring Harbor Perspectives in Biology* 7, no. 2 (2015): a021808; doi: 10.1101/cshperspect.a021808.

89 **the experience of MX:** MX's story appears in *Discover* online: Carl Zimmer, "The Brain: Look Deep Into the Mind's Eye," Discover, March 22, 2010; http://discovermagazine.com/2010/mar/23-the-brain-look-deep-into-minds-eye.

91 **Einstein described his own spatial imagery:** Jacques Hadamard, *The Psychology of Invention in the Mathematical Field*, 2nd ed. (Mineola, NY: Dover, 1954).

91 **Dyslexic mathematician Kalvis Jansons:** Kalvis M. Jansons, "A Personal View of Dyslexia and of Thought Without Language." In *Thought Without*

Language, ed. L. Weiskrantz (New York: Clarendon Press/Oxford University Press, 1988), 498–503.

91 **"what I often think of as a tactile world":** Jansons, "A Personal View of Dyslexia and of Thought Without Language." Einstein similarly described some of his mental imagery as being of a "muscular type." In Hadamard, *The Psychology of Invention in the Mathematical Field*.

Chapter 9. Trade-offs with M-Strengths

95 **researchers have found that:** Stanislas Dehaene, *Reading in the Brain: The Science and Evolution of a Human Invention* (New York: Viking, 2009).

95 **suppress the generation of its mirror image:** Dehaene, *Reading in the Brain*.

96 **Published studies have shown:** Nathlie A. Badian, "Does a Visual-Orthographic Deficit Contribute to Reading Disability?" *Annals of Dyslexia* 55, no. 1 (2005): 28–52; doi: 10.1007/s11881-005-0003-x.

96 **dyslexic designer Sebastian Bergne:** https://www.iconeye.com/icon-013-june-2004/dyslexia-icon-013-june-2004.

97 **Dyslexic biochemist Dr. Roy Daniels:** Rosalie Fink, *Why Jane and John Couldn't Read—and How They Learned: A New Look at Striving Readers* (Newark, NJ: International Reading Association, 2006).

97 **difficulties with procedural learning:** Roderick I. Nicolson and Angela J. Fawcett, *Dyslexia, Learning and the Brain* (Cambridge, MA: MIT Press, 2010).

97 **mastered more slowly:** Persistent generation of mirror-image symbols—which results from the preservation of bilateral brain processing pathways—appears to be yet another example of how the slower acquisition of mature or expert processing in many individuals with dyslexia may lead to persistence of less mature and more bilateral (or bihemispheric) brain processing.

98 **Psychologists George Hynd and Jeffrey Gilger:** Jason G. Craggs et al., "Brain Morphology and Neuropsychological Profiles in a Family Displaying Dyslexia and Superior Nonverbal Intelligence." *Cortex* 42, no. 8 (2006): 1107–18; doi: 10.1016/S0010-9452(08)70222-3.

98 **brain regions that are normally used:** These brain regions include the planum temporale, supramarginal gyrus, and angular gyrus.

98 **Einstein described the process:** Jacques Hadamard, *The Psychology of Invention in the Mathematical Field*, 2nd ed. (New York: Dover, 1954), 143.

99 **is likely one reason:** Two additional studies, one led by Vanderbilt language specialist Dr. Stephen Camarata and the other by Stanford economist Dr. Thomas Sowell, have also shown that severe late-talking is more common in children whose close family members work in analytic occupations. Many of these occupations, like engineering, scientific research, and airline piloting, are high M-strength professions. Both studies are discussed in Thomas Sowell, *The Einstein Syndrome: Bright Children Who Talk Late* (New York: Basic Books, 2002).

99 **Dr. Alison Bacon and her colleagues:** Alison M. Bacon, Simon J. Handley, and Emma L. McDonald, "Reasoning and Dyslexia: A Spatial Strategy May

Impede Reasoning with Visually Rich Information." *British Journal of Psychology* 98, pt. 1 (2007): 79–82; doi: 10.1348/000712606X103987.

Chapter 10. Growing Up with M-Strengths

107 **pioneering neurosurgeon Dr. Fred Epstein:** See Fred Epstein and Joshua Horwitz, *If I Get to Five: What Children Can Teach Us About Courage and Character* (New York: Holt Paperbacks, 2004).

Chapter 11. Key Points About M-Strengths

110 **His name was Jørn:** https://www.pritzkerprize.com/biography-jorn-utzon.

Chapter 12. What's New with M-Strengths

113 **In our research:** These 13 questions were part of a 64-question MIND strengths questionnaire that we developed by surveying over 2,200 adults using over 5,000 surveys. For those interested in more details on how the study was conducted, we've included a summary in the Appendix.

118 **In the first experiment:** Matthew Schneps, James R. Brockmole, Gerhard Sonnert, and Marc Pomplun, "History of Reading Struggles Linked to Enhanced Learning in Low Spatial Frequency Scenes." *PLoS ONE* 7, no. 4 (2012): e35724; doi: 10.1371/journal.pone.0035724.

119 **In the second experiment:** Matthew Schneps et al., "Dyslexia Linked to Visual Strengths Useful in Astronomy." In American Astronomical Society AAS Meeting #218, id.215.08. *Bulletin of the American Astronomical Society* 43 (2011).

119 **Mirela Duranovic and colleagues:** Mirela Duranovic, Mediha Dedeic, and Miroslav Gavrić, "Dyslexia and Visual-Spatial Talents." *Current Psychology* 34 (2015): 207–22; doi: 10.1007/s12144-014-9252-3.

119 **Kenneth Pugh and colleagues:** Joshua John Diehl et al., "Neural Correlates of Language and Non-Language Visuospatial Processing in Adolescents with Reading Disability." *NeuroImage* 101 (2014): 653–66; doi: 10.1016/j.neuroimage.2014.07.029.

120 **It has also more recently come to light:** Robert M. Mok and Bradley C. Love, "A Nonspatial Account of Place and Grid Cells Based on Clustering Models of Concept Learning." *Nature Communications* 10, no. 1 (2019): 5685; doi: 10.1038/s41467-019-13760-8.

122 **As Beryl recalls:** Goeff Watts, "Beryl Benacerraf: New AIUM President Gets the Picture." *Lancet* 385, no. 9973 (2015): 1065; doi: 10.1016/S0140-6736(15)60590-6.

123 **As Beryl recalled:** Watts, "Beryl Benacerraf: New AIUM President Gets the Picture."

Chapter 13. The "I" Strengths in MIND

133 **creating conceptual models that are analogous:** Nikolaus Kriegeskorte and Katherine R. Storrs, "Grid Cells for Conceptual Spaces?" *Neuron* 92, no. 2 (2016): 280–84; doi: 10.1016/j.neuron.2016.10.006.

Chapter 14. The Advantages of I-Strengths

135 **In a paper published in 1999:** John Everatt, Beverley Steffert, and Ian Smythe, "An Eye for the Unusual: Creative Thinking in Dyslexics." *Dyslexia* 5, no. 1 (1999): 28–46; doi: 10.1002/(SICI)1099-0909(199903)5:<28::AID-DYS126>3.0.CO;2-K.

137 **Ambiguities in word meanings:** We owe these interesting facts to Jeff Gray at the Gray-Area website (www.gray-area.org/Research/Ambig/).

138 **the Gaia hypothesis:** James Lovelock, *The Revenge of Gaia* (New York: Basic Books, 2006).

143 **T. R. Miles and colleagues:** T. R. Miles, Guillaume Thierry, Judith Roberts, and Josie Schiffeldrin, "Verbatim and Gist Recall of Sentences by Dyslexic and Non-Dyslexic Adults." *Dyslexia* 12, no. 3 (2006): 177–94; doi: 10.1002/dys.320.

144 **this kind of upside surprise:** Remarkably, we even see this kind of "upside surprise" in story comprehension in some of the children diagnosed with specific language impairment (SLI), which shares many of the processing features of dyslexia but is associated with more severe difficulties in comprehending language. These children typically have difficulties comprehending all but the shortest and most transparent sentences. When given a story with enough context and redundancy, these children often comprehend far better than expected. We've seen children score as much as three standard deviations higher on their oral story comprehension than on their vocabulary and single-sentence comprehension. This is likely due to their strengths in gist detection and top-down context-based processing.

Chapter 15. Trade-offs with I-Strengths

148 **Sometimes these substitutions involve:** Also called *malapropisms*.

149 **Conceptual substitutions like these:** Thomas G. West has an interesting and insightful discussion of the phenomenon of paralexia in West, *In the Mind's Eye: Creative Visual Thinkers, Gifted Dyslexics, and the Rise of Visual Technologies*, 2nd ed. (Amherst, NY: Prometheus Books, 2009), 43ff.

150 **one study looking at Harvard students:** Shelley H. Carson, Jordan B. Peterson, and Daniel M. Higgins, "Decreased Latent Inhibition Is Associated with Increased Creative Achievement in High-Functioning Individuals." *Journal of Personality and Social Psychology* 85, no. 3 (2003): 499–506; doi: 10.1037/0022-3514.85.3.499.

Chapter 16. Brains Without Borders

155 **"maximized my likelihood of failing":** In reading, too, Douglas developed strategies. "When I could no longer get away with manipulating people, I built a bunch of tricks to try to get through reading. . . . It's mostly just modified skimming methodologies where you just sort of tag stuff as you go by, because there was no way I was going to be able to read things in any detail. But I could skim, and I would mark something like 'This might be interesting later,' and then I would skim the thing over and over again, but I'd try not to think of

it as reading, because if I thought of it as reading, I'd get all worried about failing and how hard it was and I'd work myself into a frenzy, and that was not helpful. But by skimming over and over again and making progressively more organized marks, I could get the key elements out of an article or a paper or a chapter."

158 **"there's lots of different views on problems":** This belief is beautifully exemplified by one of his projects at Google: "Imagine that people could ask questions of the world around them and get back answers that don't entirely match their perspective. How terrific would it be if it were possible for all of us to read what the Arabic newspapers were saying about our operations in the Middle East. How good would it be for the world if the democratization of information got to the place where consumers could see their own perspectives, the perspectives of those they trust and the perspectives of people who disagree with them all together and compare them." Abbie Lundberg, "IT's Third Epoch . . . and Running IT at Google." *CIO*, October 8, 2007; https://www.cio.com/article/274565/consumer-technology-it-s-third-epoch-and-running-it-at-google.html.

Chapter 18. What's New with I-Strengths

168 **there were fourteen studies in one meta-analysis:** Nadyanna M. Majeed, Andree Hartanto, and Jacinth J. X. Tan, "Developmental Dyslexia and Creativity: A Meta-Analysis." *Dyslexia* 27, no. 2 (2021): 187–203; doi: 10.1002/dys.1677.

168 **twenty in the other:** Florina Erbeli, Peng Peng, and Marianne Rice, "No Evidence of Creative Benefit Accompanying Dyslexia: A Meta-Analysis." *Journal of Learning Disabilities* 55, no. 3 (2022): 242–53; doi: 10.1177/00222194211010350.

169 **generate novel metaphors:** Anat Kasirer and Nira Mashal, "Comprehension and Generation of Metaphoric Language in Children, Adolescents, and Adults with Dyslexia." *Dyslexia* 23, no. 2 (2017): 99–118; doi: 10.1002/dys.1550.

171 **excel at making remote connections:** Kristoffer C. Aberg, Kimberly C. Doell, and Sophie Schwartz, "The 'Creative Right Brain' Revisited: Individual Creativity and Associative Priming in the Right Hemisphere Relate to Hemispheric Asymmetries in Reward Brain Function." *Cerebral Cortex* 27, no. 1 (2017): 4946–59; doi: 10.1093/cercor/bhw288.

171 **These studies have shown:** Kevin P. Madore et al., "Neural Mechanisms of Episodic Retrieval Support Divergent Creative Thinking." *Cerebral Cortex* 29, no. 1 (2019): 150–66; doi: 10.1093/cercor/bhx312. Roger E. Beaty et al., "Default Network Contributions to Episodic and Semantic Processing During Divergent Creative Thinking: A Representational Similarity Analysis." *NeuroImage* 209 (2020): 116499; doi: 10.1016/j.neuroimage.2019.116499.

173 **"I just couldn't do the SATs":** The Scholastic Aptitude Test, a standardized test required for admission to many colleges.

173 **"I wasn't college material":** "Portrait of a Dyslexic Artist, Who Transforms Neurons Into 'Butterflies,'" *PBS NewsHour*, April 16, 2014; https://www.pbs.org/newshour/science/neuroscience-art-brain.

174 **The magazine *Penn Today*:** "Reimagining Scientific Discovery Through the Lens of an Artist," *Penn Today*, September 17, 2021; https://penntoday.upenn.edu/news/reimagining-scientific-discovery-through-lens-artist.

175 **When Chuck was a boy:** The material for this profile was drawn from Chuck's autobiography: Charles Harrison, *A Life's Design: The Life and Work of Industrial Designer Charles Harrison* (Chicago: Ibis Design, 2005).

Chapter 19. Narrative Reasoning: The N-Strengths in MIND

181 **"School was torture":** Both this and all other quotes attributed to Anne Rice have, unless otherwise stated, been taken from her autobiography, *Called Out of Darkness: A Spiritual Confession* (New York: Alfred A. Knopf, 2008).

Chapter 20. The Advantages of N-Strengths

189 **"We left our house":** Anne Rice, *Called Out of Darkness: A Spiritual Confession* (New York: Alfred A. Knopf, 2008), 17–18.

189 **Dr. Demis Hassabis:** Demis has a curriculum vitae that sounds like it was dreamed up by Stan Lee as the backstory for a superhero. While he doesn't sling webs or turn green and bulk up when he gets angry, he was a chess master at age thirteen, won the Pentamind World Championship at the Mind Sports Olympiad a record five times, and became a successful video game designer at age seventeen. He also earned a double first-class degree in computer science from Cambridge and started a successful video game production company with sixty-five employees—all by the time he'd reached his mid-twenties. After successfully selling his company, he decided to combine his interests in imagination, creativity, and artificial intelligence by pursuing a PhD in cognitive neurosciences at University College London. Since the first edition of this book was published Demis cofounded the artificial intelligence company DeepMind, became its CEO, and has been named a Fellow of the Royal Society.

190 **with his colleague Dr. Eleanor Maguire:** Demis Hassabis, Dharshan Kumaran, Seralynne D. Vann, and Eleanor A. Maguire, "Patients with Hippocampal Amnesia Cannot Imagine New Experiences." *Proceedings of the National Academy of Sciences of the USA* 104, no. 5 (2007): 1726–31; doi: 10.1073/pnas.0610561104. See also Demis Hassabis and Eleanor A. Maguire, "Deconstructing Episodic Memory with Construction." *Trends in Cognitive Science* 11, no. 7 (2007): 299–306; doi: 10.1016/j.tics.2007.05.001.

196 **schemas appear to be generated through a process:** Louis Renoult, Muireann Irish, Morris Moscovitch, and Michael D. Rugg, "From Knowing to Remembering: The Semantic-Episodic Distinction." *Trends in Cognitive Science* 23, no. 12 (2019): 1041–57; doi: 10.1016/j.tics.2019.09.008.

198 **a picture called the "Cookie Thief":** From a test called the Boston Diagnostic Aphasia Examination. H. Goodglass and E. Kaplan, *Boston Diagnostic Aphasia Examination*, 2nd ed. (Philadelphia: Lea and Febiger, 1983).

Chapter 23. Key Points About N-Strengths
215 **including the outstanding *Unlimited Memory***: Kevin Horsley, *Unlimited Memory: How to Use Advanced Learning Strategies to Learn Faster, Remember More, and Be More Productive* (TCK Publishing.com, 2016).

Chapter 24. What's New with N-Strengths
220 **Incidental memories are often stored:** They may also be stored as more vague or "fuzzy" gist-trace memories. Michał Obidziński and Marek Nieznański, "False Memory for Orthographically Versus Semantically Similar Words in Adolescents with Dyslexia: A Fuzzy-Trace Theory Perspective." *Annals of Dyslexia* 67, no. 3 (2017): 318–32; doi: 10.1007/s11881-017-0146-6.

220 **Michael Ullman of Georgetown University:** Martina Hedenius et al., "Enhanced Recognition Memory After Incidental Encoding in Children with Developmental Dyslexia." *PLoS ONE* 8, no. 12 (2013): 0063998; doi: 10.1371/journal.pone.0063998.

222 **discovered entirely by accident:** Marcus E. Raichle et al., "A Default Mode of Brain Function." *Proceedings of the National Academy of Sciences of the USA* 98, no. 2 (2001): 676–82; doi: 10.1073/pnas.98.2.676.

223 **Soon, however, it became apparent:** Roger E. Beaty, "The Creative Brain." *Cerebrum 2020* (2020): cer-02-20; PMID: 32206175.

223 **plays a vital role:** Pam Hruska et al., "Hemispheric Activation Differences in Novice and Expert Clinicians During Clinical Decision Making." *Advances in Health Science Education* 21, no. 5 (2016): 921–33; doi: 10.1007/s10459-015-9648-3.

224 **As creativity researcher Kevin Madore has written:** Kevin P. Madore et al., "Neural Mechanisms of Episodic Retrieval Support Divergent Creative Thinking." *Cerebral Cortex* 29, no. 1 (2019): 150–66; doi: 10.1093/cercor/bhx312.

Chapter 25. The "D" Strengths in MIND
240 **"test and integrate":** Sarah Andrews, "Spatial Thinking with a Difference: An Unorthodox Treatise on the Mind of the Geologist." *AEG News* 45, no. 4 (2002), and 46, nos. 1–3 (2003).

240 **astonished her teachers:** In contrast to her poor verbal performance, Sarah excelled on the math portion of the SAT—despite being a C student in math class—prompting her math teacher to ask her, "Where have you been hiding this?" Sarah explained that the difference was entirely due to the fact that the SAT did not require her to show her work, which eliminated her problems showing work and removed any penalty for her original way of doing math. Like many of the individuals we've mentioned in previous chapters, Sarah had difficulty memorizing and following the standard math formulas and procedures, so she created her own and did most of her work in her head. This led to conflicts with her teachers. "My goal was, 'Let's get the right answer,' but theirs was, 'Let's do it the right way.'"

241 **Lysbeth was dyslexic:** In a fascinating twist on this story, rather than work as a geologist, Sarah's aunt Lysbeth taught grade school and became a specialist in teaching what she termed "reluctant readers."

Chapter 26. The Advantages of D-Strengths

248 ***Dinosaurs Under the Big Sky:*** Jack Horner, *Dinosaurs Under the Big Sky* (Missoula, MT: Mountain Press, 2001).

249 **Sarah further described this process:** Sarah Andrews, "Spatial Thinking with a Difference: An Unorthodox Treatise on the Mind of the Geologist." *AEG News* 45, no. 4 (2002), and 46, nos. 1–3 (2003).

250 **Sarah points out why:** Andrews, "Spatial Thinking with a Difference."

Chapter 27. Trade-offs with D-Strengths

253 **Sarah Andrews echoed this description:** Sarah Andrews, "Spatial Thinking with a Difference: An Unorthodox Treatise on the Mind of the Geologist." *AEG News* 45, no. 4 (2002), and 46, nos. 1–3 (2003).

253 **Sarah later wrote of this episode:** Andrews, "Spatial Thinking with a Difference."

255 **that make up the process of insight:** Mark Jung-Beeman et al., "Neural Activity When People Solve Verbal Problems with Insight." *PLoS Biology* 2 (2004): 500–510; doi: 10.171/journal.pbio.0020097.

Chapter 28. D-Strengths in Business

263 **Julie eventually conducted:** Dr. Logan reported that the incidence of dyslexia is 20 percent among entrepreneurs in the United Kingdom, where the population incidence of dyslexia is estimated at 4 percent, and 35 percent of entrepreneurs in the United States, where the population incidence is around 10 to 15 percent.

264 **their labor standoff continued:** Glenn Bailey also gave us a great example of the way that personal relationships can greatly affect worker satisfaction and performance. "When we ran our first water company, we had a great relationship with our team, and we had virtually no work-related injury claims, despite the fact that we hand-delivered all these huge five-gallon bottles of purified water. When we sold the company, the people who bought us out were all about cash and bottom line, and they got rid of the Ping-Pong table and the barbecue and they got unionized, and their claims went through the roof. As a result, they became number one in worker injury claims in British Columbia."

Chapter 29. Key Points About D-Strengths

267 **one of the top-selling novelists of his era:** Vince's character Mitch Rapp was so popular with his many fans that the series of books about him has continued on for seven more bestsellers under author Kyle Mills.

267 **"he just doesn't test well":** Robert J. Bidinotto, "An Interview with Vince Flynn (Part 1)," https://www.bidinotto.com/2011/10/an-interview-with-vince-flynn-part-1/.

Chapter 30. What's New with D-Strengths

277 **additional higher-cognitive functions:** Roger E. Beaty, "The Creative Brain." *Cerebrum 2020* (2020): cer-02-20; PMID: 32206175.

277 **counterfactual reasoning:** Nicole Van Hoeck, Patrick D. Watson, and Aron K. Barbey, "Cognitive Neuroscience of Human Counterfactual Reasoning." *Frontiers in Human Neuroscience* 9 (2015): 420; doi: 10.3389/fnhum.2015.00420.

277 **perspective shifting:** Van Hoeck, Watson, and Barbey, "Cognitive Neuroscience of Human Counterfactual Reasoning."

277 **storytelling and creative writing:** Qunlin Chen, Roger E. Beaty, and Jiang Qiu, "Mapping the Artistic Brain: Common and Distinct Neural Activations Associated with Musical, Drawing and Literary Creativity." *Human Brain Mapping* 41, no. 12 (2020): 3403–19; doi: 10.1002/hbm.25025.

277 **creation of visual arts:** Chen, Beaty, and Qui, "Mapping the Artistic Brain."

277 **complex problem-solving:** Pam Hruska et al., "Hemispheric Activation Differences in Novice and Expert Clinicians During Clinical Decision Making." *Advances in Health Science Education* 21, no. 5 (2016): 921–33; doi: 10.1007/s10459-015-9648-3.

277 **generation of vivid mental imagery:** Vishnu Sreekumar et al., "The Experience of Vivid Autobiographical Reminiscence Is Supported by Subjective Content Representations in the Precuneus." *Scientific Reports* 8, no. 1 (2018): 14899; doi: 10.1038/s41598-018-32879-0.

277 **and even reading:** The role of the default mode network in reading is especially interesting and relevant for individuals with dyslexia. Researchers have shown that dyslexic readers tend to activate the default mode network more often during reading than do skilled readers. See, e.g., Emily S. Finn et al., "Disruption of Functional Networks in Dyslexia: A Whole-Brain, Data-Driven Analysis of Connectivity." *Biological Psychiatry* 76, no. 5 (2014): 397–404; doi: 10.1016/j.biopsych.2013.08.031. This activation has been interpreted as an attempt to compensate for their dysfunction of more efficient reading pathways, and to some extent this is likely true. Recall that the default mode network is activated by tasks that are especially difficult or unfamiliar or that draw heavily upon context for interpretation. Reading is often all of these for dyslexic readers.

However, the default mode network may actually be more helpful to good reading than previously suspected. Professor of comparative literature Karin Kukkonen of the University of Oslo has recently described the essential role that mind-wandering plays for skilled readers when they are reading literary texts. See, e.g., Regina E. Fabry and Karin Kukkonen, "Reconsidering the Mind-Wandering Reader: Predictive Processing, Probability Designs, and Enculturation." *Frontiers in Psychology* 9 (2019): 2648; doi: 10.3389/fpsyg.2018.02648. She writes: "What brings texts to life when reading often looks very much like mind-wandering." This is true, she noted, particularly at those points when the reader must "share in filling the blanks which written texts inevitably leave." This value of the default mode network for reading may explain why so many dyslexic individuals actually love to read—whether when possible by conventional "eye reading" or by listening to recordings or text-to-speech audio. It may also explain why they are often especially perceptive readers, even though they struggle with the mechanical act of reading itself.

279 **Jeff Hawkins:** Jeff Hawkins, *A Thousand Brains: A New Theory of Intelligence* (New York: Basic Books, 2021), 80.

279 **the state of *surprise*:** Talia Brandman, Rafael Malach, and Erez Simony, "The Surprising Role of the Default Mode Network in Naturalistic Perception." *Communications Biology* 4, no. 1 (2021): 79; doi: 10.1038/s42003-020-01602-z.

280 **error-driven learning:** Brandman, Malach, and Simony, "The Surprising Role of the Default Mode Network in Naturalistic Perception."

284 **decreases the activity of their default mode network:** Hikaru Takeuchi et al., "Failing to Deactivate: The Association Between Brain Activity During a Working Memory Task and Creativity." *NeuroImage* 55, no. 2 (2011): 681–87; doi: 10.1016/j.neuroimage.2010.11.052.

284 **does not occur:** Takeuchi et al., "Failing to Deactivate: The Association Between Brain Activity During a Working Memory Task and Creativity."

Chapter 31. Exploratory Strengths of the Dyslexic Mind

297 **has published her own theory:** Helen Taylor and Martin David Vestergaard, "Developmental Dyslexia: Disorder or Specialization in Exploration?" *Frontiers in Psychology* 13 (2022); doi: 10.3389/fpsyg.2022.889245.

303 **"This trade-off arises":** Taylor and Vestergaard, "Developmental Dyslexia: Disorder or Specialization in Exploration?"

303 **One well-studied example is the business environment:** See, e.g., James G. Marsh, "Exploration and Exploitation in Organizational Learning." *Organization Science* 2, no. 1 (1991): 71–87; https://www.jstor.org/stable/2634940.

307 **his fascinating autobiography, *Into the Deep*:** Robert Ballard and Christopher Drew, *Into the Deep: A Memoir from the Man Who Found Titanic* (Washington, D.C.: National Geographic, 2021).

307 **"I've always felt insecure":** Ballard and Drew, *Into the Deep*, 285.

309 **"I've learned to embrace failure as the greatest teacher":** Ballard and Drew, *Into the Deep*, 305, 12.

309 **"I could overcome any challenge":** Ballard and Drew, *Into the Deep*, 285.

312 **"That's how much it mirrored":** Ballard and Drew, *Into the Deep*, 285–86.

Chapter 32. Educating Dyslexic MINDs

317 **begun in students before age nine:** Maureen W. Lovett et al., "Early Intervention for Children at Risk for Reading Disabilities: The Impact of Grade at Intervention and Individual Differences on Intervention Outcomes." *Journal of Educational Psychology* 109, no. 7 (2017): 889–914; doi: 10.1037/edu000181.

324 **The following egregious example:** D. Hensley, "Children Who Do Not Learn to Read, Write by 3rd Grade Face Tough Future." *Wichita Eagle*, June 5, 2019.

328 **enhanced emotional reactivity:** Eleanor R. Palser et al., "Children with Developmental Dyslexia Show Elevated Parasympathetic Nervous System Activity at Rest and Greater Cardiac Deceleration During an Empathy Task." *Biological Psychology* 166 (2021): 108203; doi: 10.1016/j.biopsycho.2021.108203. Virginia E. Sturm et al., "Enhanced Visceromotor Emotional Reactivity in Dyslexia and Its Relation to Salience Network Connectivity." *Cortex* 134 (2021): 278–95; doi: 10.1016/j.cortex.2020.10.022.

Chapter 33. Dyslexic MINDs at Work

351 **in the book *Moneyball*:** Michael Lewis, *Moneyball: The Art of Winning an Unfair Game* (New York: W. W. Norton, 2003).

Epilogue

378 **filled with self-doubt:** In a survey done through our Dyslexic Advantage nonprofit, roughly 90 percent of the parents of over 200 dyslexic children reported seeing visible loss of self-esteem in their child within three months of beginning to struggle in school, and many of these children eventually experienced even more serious emotional and behavioral challenges.

Index

A
abstract memory. *See* semantic (abstract/impersonal) memory
abstract reasoning, 16–17, 120, 192–195, 202–206, 250–251. *See also* semantic (abstract/impersonal) memory
abstract verbal definitions, 192, 225–226
ADHD, 68–69, 256–257
advantages of dyslexia, xvi–xix, 4–5, 6. *See also* dynamic reasoning; exploratory strengths; interconnected reasoning; material reasoning; MIND strengths survey; narrative reasoning; work and occupational successes
advocacy for dyslexic people, 366–367, 369–371, 373–376, 377–380
alternative uses task, 135
Alternative Uses Test, 169
Alvin (deep-sea research vehicle), 309
ambiguities, 137–138

Amscan, 22–23
Andrews, Sarah, 239–243, 246–250, 253, 407–408n240
aphantasia, 90
Aragorn metaphor, 380–381
Aristotle, 280–281
Asperger's, 104
assessment, in dyslexic learning loop, 323
assistive technology, 233, 357–358, 370
associations and similarities (perceiving relationships), 17, 134–139, 146–150, 166, 170–171
associative theory of creativity, 170–171
attentional immaturity, 256–257
attention skills, 68–69
attitude
  entrepreneurial success and, 263
  for learning, 280–285, 321–323, 337, 366
  for modeling of dyslexic self, 364–366
Attree, Elizabeth, 82–83, 84, 220

audiobooks, 370
auditory-verbal working memory, 51–52
autism, 44, 47–48, 104
automaticity, 38, 49–52, 54–55, 56–57
automatization deficit hypothesis, 55
axons, 44, 47

B
backcasting, 349
Bacon, Alison, 99–100
Bailey, Glenn, 89, 259–261, 408n264
Ballard, Robert "Bob," 306–314
Banks, Larry, 230–235
Bannatyne, Alexander, 401n86
baseball
  predictions based on batting averages, 250–251
  talent pool comparison, 20–21, 351
Bassett, Dani, 174–175
Beeman, Mark, 38–39, 254–255, 256–257, 398n40
Benacerraf, Beryl, 121–124
Benedetto, Ida, 350
Benton, Robert, 184
Bergne, Sebastian, 96
big-picture (global) thinking, 142–145, 152–153, 159, 167
big-P Predictions, 278
binary distribution model of dyslexia, 25
Bishop, Elizabeth, 230
Bookshare.org, 370
bottom-up teaching, 152–153
Boy Scouting, 129
brain function, 33–48
  at behavioral level, 33, 38
  at cognitive level, 33, 35, 49–59, 279–280. *See also* conscious mental resources; *specific types of memory*
  conclusion, 48

  at network level, 33, 35, 44–48, 55–57, 299n47, 399n48
  overview, 33–34
  at structural level, 34–41, 45, 95, 98, 171, 196, 401n98. *See also specific brain structures*
  3-D modeling and, 88–89
Branson, Richard, 264
Brooks, Rodney, 287–288

C
Called Out of Darkness (Rice), 188–189, 406n181
Camarata, Stephen, 402n99
Canadian Springs water company, 261
career options. *See* work and occupational successes
Carroll, Lewis, 189
Casanova, Manuel "Manny," 42, 44–45, 47, 399n48
case-based knowledge, 205
cerebellum, 54
Charlton, Blake, 207–210, 221–222, 325
charter schools, 326–327
Cheever, John, 230
cognitive search, 304
Cog robot project, 288
columns, 42–43. *See also* cortex of brain
Comforts of the Abyss (Schultz), 230
compact disc system, 19
complex systems theory, 302–303
conscious compensation, 51, 305
conscious mental resources, 49–59
  automaticity and, 49–52, 54–55, 56–57
  conclusion, 58–59
  declarative memory and, 57–58. *See also* declarative (long-term) memory
  neural noise and delayed network formation, 53–56

overview, 49–50
procedural learning and
  memory, 50–53, 57–58. See also
  procedural learning and memory
Copernican revolution, 6, 379–380
Corcoran, Barbara, 5
cortex of brain
  columns of, 42–43
  minicolumns of, 43–48, 55,
    171, 305
covert orienting, 64n
Cox, Allan, 310
creativity
  associative theory of, 170–171
  defined, 168
  development and, 256–257
  distractions and, 150
  divergent thinking and, 171–172,
    224–225, 304, 343–344
  D-strengths and, 284
  exploratory strengths and, 304
  imagination and, 255
  I-strengths and, 168–170,
    174–175, 343–346
  M-strengths and, 341–343
  N-strengths and, 180–181, 195,
    224–225, 230, 255–257,
    346–348
  visual-spatial, 82–85, 135–136
Critchley, Macdonald, 87
cube root rule, 52

D
Daniels, Roy, 97
daydreaming, 17, 253, 255, 258,
  409–410n277
decisions, in dyslexic learning
  loop, 322
declarative (long-term) memory
  advantages of, 219–222
  creation and processing of,
    195–197
  overview, 57–58, 186–187
default mode network
  D-strengths and, 244, 255,
    276–277, 409–410n277

error-driven learning and, 284
N-strengths and, 222–226
Plan B skills and, 349
surprise state and, 280, 284
Delahaye, Valerie, 87
delayed neural commitment, 57
development
  creativity and, 256–257
  late bloomers, 68, 106–107,
    239–241, 327, 365
  late-talking, 99, 402–403n99
  M-strengths and, 102–108,
    109–110
  working memory and, 99–100,
    104–106, 169–170, 365
developmental language
  disorder, 62
Didion, Joan, 230
Dinosaurs Under the Big Sky
  (Horner), 248
distractibility, 119, 150
double deficit, 63
Drennan, Catherine, 75–80
D-strengths. See dynamic
  reasoning
Duranovic, Mirela, 119
dynamic reasoning (D-strengths),
  239–293
  advantages of, 246–251
  entrepreneurial success and,
    259–265
  key points about, 266–267
  new research relevant to,
    276–285
  overview, xix, 246–248
  people with, 239–243, 259–261,
    267–272, 285–293
  prediction and, 243–245
  as subset of N-strengths,
    244–245
  survey on, 273–276, 394–395
  trade-offs with, 252–258
  value of mistakes and,
    279–285
  in work and workplace, 341,
    348–350

dyslexia

advantages, xvi–xix, 4–5, 6. *See also d*ynamic reasoning; exploratory strengths; *i*nterconnected reasoning; *m*aterial reasoning; MIND strengths survey; *n*arrative reasoning

career options, 5, 339. *See also* work and occupational successes

challenging factors for, 5–6, 60–72. *See also* reading and spelling challenges

demographics, 8, 86, 397n8, 408n263

education principles for, 315–338. *See also* educating dyslexic minds

future directions, 6, 377–381

modeling of, 22–29. *See also* modeling dyslexia

narrow view of, xv–xvi, 7–10, 12–15

new view of (general), xvii–xviii, 3–7, 11–19

new view of talent-dyslexia relationship, 19–21

protective factors in, 69–71

resources, 378–379

scientific dimensions of, 33–48, 49–59. *See also* brain function

self-concept, 361–376. *See also* modeling the dyslexic self

dyslexic learning loop, 322–323

dyslexic processing style, 4–6

E

"ear learning," 370

Echavarria, Amparo, 368–376

Echavarria, Valerie, 368–376

Eden, Guinevere, 36

educating dyslexic minds, 315–338
  in action, 328–338
  Ballard on, 313–314
  big-picture principles for, 317–323
  equality in education, 313, 325–328, 375–376
  new approach to, 316–317, 323–327, 378–380, 404–405n155
  overview, 315–316
  reading and learning approaches, 323–325

Einstein, Albert, 91, 98, 99, 106–107

emergent properties, 46

EMI Recorded Music, 157, 201

entrepreneurial success
  demographics, 408n263
  D-strengths for, 259–265
  people with, xv, 259–261
  research on, 262–265

environmental stimuli, 150

episodic (personal) memory
  complementary cognition theory on, 305
  creation and processing of, 171–172, 195–201
  default mode network, 222–226, 244, 255, 276–277
  D-strengths and, 244–245, 255, 276–277, 281–282
  for educating dyslexic minds, 317–320, 326
  functions of, 188–191
  incidental learning/memories and, 220, 281–282
  I-strengths and, 171–172
  key points about, 211–212
  N-strengths and, 186–201, 222–226
  overview, 186–188
  trade-offs with, 16–17, 120, 192–195, 202–206, 225–226, 250–251

episodic construction, 211, 244–245, 247–251, 346–347

episodic retrieval orientation, 172, 224–225

episodic scene construction, 190–191, 224–225, 250, 277, 319

episodic simulation, 190–191, 244–245, 248–250, 266
episodic specificity induction, 171–172
Epstein, Fred, 107–108
error-driven learning, 279–285, 321–323, 337, 366
Everatt, John, 135
"The Everest of Memory Tests," 215
Evernote, 357
Evers, Tony, 359
executive functioning skills, 68–69
experience-based learning, 283, 347–348, 365, 373
experiential memory. *See* episodic (personal) memory
explicit learning, 51, 102, 104
explicit processing, 57–58
exploration strategy, 298–306
exploratory process of learning, 321–323
exploratory strengths
  complementary cognition theory, 298–306
  educating dyslexic minds through, 321–323
  overview, 297
  people with, 306–314

F
Failure (Schultz), 229
failure, embracing, 309, 337, 365–366
Fawcett, Angela, 49–50, 52, 55, 56–58
fixations, 54, 64
Flynn, Vince, 184, 267–272, 325, 408n267
focused attention, 150, 256–257, 282–284
Ford, Chris, 352–360
Ford, Richard, 184
forecasting. *See* prediction
free association, 151
fresh thinking, 343–344
Fundamentals of English, 269

G
Gaia hypothesis, 138–139
geological reasoning, 241–243, 247–250
Gershoni, Gil, 327
Geschwind, Norman, 86, 87
Gilger, Jeffrey, 98
gist and gist detection, 34, 143–145, 344, 398n40, 399n48, 404n144
gist tracings, 196–197, 407n220
global thinking, 142–145, 152–153, 159, 167
Goodenough, John, 5
Google, 157, 290, 405n158
Google X, 290
Graduate Record Examination (GRE), 78, 337
Greider, Carol, 5, 380
grid cells, 88–89, 120–121, 223, 277
grit, 365–366, 371

H
Harrison, Chuck, 5, 175–178
Hassabis, Demis, 189–191, 193–195, 406n189
Hawkins, Jeff, 43, 279
Hebb's rule, 53–54
hemispheres of the brain
  mirror-image views and, 95
  processing style and, 34–41, 45, 171, 398n40, 399n48
  semantic memory and, 196
hippocampus, 88–89, 88n, 90, 120–121, 193–195, 280
Hoeft, Fumiko, 70
Horner, John "Jack," 140–142, 162, 248, 341, 380
Horsley, Kevin, 212–215, 221–222, 319, 325
How to Spell Five Words a Day, 269
human brain. *See* brain function
Hynd, George, 98

I

IDA (International Dyslexia Association), 9, 210, 370
IDEA learning loop, 322–323
imagination
  defined, 255
  in dyslexic learning loop, 322
  episodic memory and, 244–245
impersonal memory. See semantic (abstract/impersonal) memory
implicit learning, 50–51, 281–282
implicit processing, 57–58
impossible figures, 84–85, 119–120
incidental memory, 219–221, 281–282, 407n220
industrial design, 176–178
inner dialogue, 333–334
insight-based problem-solving, 99, 254–258, 266, 304
instincts and intuition, 156, 230, 264
*i*nterconnected reasoning (I-strengths), 127–180
  advantages of, 134–145
  global thinking, 142–145, 152–153, 159, 167
  insight and, 254
  key points about, 159–163
  new research on, 167–172
  overview, xix, 132–133
  people with, 127–132, 154–158, 160–163, 172–178
  in perceiving relationships, 17, 134–139, 146–150, 166, 170–171
  in perspective shifting, 139–142, 151–152, 155–158, 167
  survey on, 164–167, 390–391
  trade-offs with, 146–153, 160
  in work and workplace, 341, 343–346
interdisciplinary mindset, 139–140, 310–311
International Dyslexia Association (IDA), 9, 210, 370

Interview with the Vampire (Rice), 183
In the Mind's Eye (West), 36
Into the Deep (Ballard), 307
intuition and instincts, 156, 230, 264
Iron Man (film), 330
Irving, John, 184
Ive, Jony, 5

J

Jackson, Lesley, 87
Jansons, Kalvis, 91
jobs. See work and occupational successes
John, Daymond, 5
Johnston, Don, 328
Jones, Johnpaul, 5

K

Kamen, Dean, 5
Kamen, Rebecca, 172–175
Károlyi, Catya von, 84, 119
Keeney, Bill, 370
Kenyon, Sherrilyn, 184
Koehl, Mimi, 160–163
Kukkonen, Karin, 409–410n277
Kurzweil Personal Reader, 233

L

language processing, 38–40, 62, 71–72, 97–101, 109, 404n144
La Plante, Lynda, 184
late bloomers, 68, 106–107, 239–241, 327, 365
latent inhibition, 150, 256
Laws, John Muir "Jack," 127–132, 138, 193, 283, 325, 328
Lean, David, 122
Learning Ally, 370
Lee, Spike, 5, 233
left hemisphere of brain, 34–40, 398n40, 399n48. See also hemispheres of the brain
Lennon, John, 5
Leonardo da Vinci, 96

letter reversals, 95–97
Lewis, Michael, 351
likeness and associations, 17, 146–150, 166, 170–171
Logan, Julie, xv, 262–263, 408n263
long-term memory. *See* declarative (long-term) memory
The Lord of the Rings (Tolkien), 380
Lovelock, James, 138–139, 140
low-pass filtering, 118
Ludwig, Martha, 78
Lundberg, Ingvar, 86
Lydon-Staley, David, 175

M
MacGregor, Kinley (Sherrilyn Kenyon), 184
magnet schools, 326–327
Maguire, Eleanor, 190
Mailer, Norman, 230
material reasoning (M-strengths), 75–124
  cognitive basis of, 88–93
  developmental issues, 102–108, 109–110
  key points about, 109–112
  new research on, 117–121
  overview, xix
  people with, 75–81, 84–85, 102–106, 110–112, 121–124
  real-world worth of, 85–87
  survey on, 113–117, 388–389
  symbol struggles and, 94–97, 109
  3-D spatial reasoning abilities, 80–81, 109
  trade-offs with, 94–101, 109
  in visual-spatial creativity, 82–85, 135–136
  in work and workplace, 340–343
math challenges
  abstract reasoning and, 250
  imagery for, 91
  insight-based problem-solving and, 257–258, 261
  memorization difficulties, 13, 14, 66, 91, 100

new view on educational approaches, 332–334
  storytelling for, 155, 156, 200
Matsuoka, Yoky, 5, 285–293, 341
Max (child with M-strengths), 102–106
McComas, David, 380
McGregor, Gary, 328–338, 339, 342–343
McGuire, Al, 268
Mednick, Sarnoff, 170–171
Memorial Day (Flynn), 271
memory. *See* declarative (long-term) memory; episodic (personal) memory; procedural learning and memory; rote memory; semantic (abstract/impersonal) memory; short-term memory; working memory
mental simulation, 252–254, 266, 311
Merrill, Douglas, 154–158, 200–201, 252–253, 404–405n155, 405n158
meta-analysis, 168–169
metacognition, 363–364
microcircuitry of the brain, 33, 42–48, 53–56, 159, 171
Miles, T. R., 143–144
MIND strengths
  D-strengths. *See d*ynamic reasoning
  I-strengths. *See i*nterconnected reasoning
  M-strengths. *See m*aterial reasoning
  N-strengths. *See n*arrative reasoning
  at work. *See* work and occupational successes
MIND strengths survey
  about, 387–388, 403n113
  D-strengths, 273–276, 394–395
  I-strengths, 164–167, 390–391
  M-strengths, 113–117, 388–389
  N-strengths, 216–219, 392–393

mind-wandering, 17, 253, 255, 258, 409–410n277
minicolumns (cortex), 43–48, 55, 171, 305, 399n48
mirror reversals, 94–97, 402n97
mis-predictions, 279
mistakes, value of, 279–285
"MIT disease," 87
mnemonic strategy, 319
modeling dyslexia
　in action, 22–24
　mistaken models, 25–26
　in multiple dimensions, 26–29
　overview, 24–25
modeling the dyslexic self, 361–376
　attitude and, 364–366
　embracing of differences, 362–364, 411n378
　family and, 368–376
　overview, 361–362
　personal story, 368–376
　supportive environment for, 327–328, 366–368
modular systems, 46
Moneyball (Lewis), 351
Morgan, W. Pringle, 8
Moser, Edvard, 88
Moser, May-Britt, 88
mountain expedition model, 27–29, 120–121
Muir, John, 127, 130, 132
multidimensional model of dyslexia, 24–25, 26–29
"multiple specialists," 139–140
My Dyslexia (Schultz), 229

N
naming speed, 62–63
narrative reasoning (N-strengths), 181–235
　advantages of, 186–201
　creative writing and, 183–185
　D-strengths as subset of, 244–245. *See also* dynamic reasoning
　episodic memory and, 188–191, 195–197
　key points about, 211–212
　new research on, 219–222
　new research relevant to, 222–226
　overview, xix
　people with, 181–185, 207–210, 212–215, 226–235
　scene-based vs. abstract knowledge, 192–195
　survey on, 216–219, 392–393
　thinking in stories, 198–201
　trade-offs with, 202–206
　in work and workplace, 341, 346–348
National Institute of Child Health and Human Development (NICHD), 9
Negroponte, Nicholas, 87
Nest thermostat, 290
networked thinking. *See interconnected reasoning*
neurobotics, 289, 290
neurons, 33, 42–48
NICHD (National Institute of Child Health and Human Development), 9
Nicolson, Roderick "Rod," 49–50, 54, 55, 56–58
nonspatial concepts, 120–121
nonverbal reasoning, 97–101
normal distribution model of dyslexia, 25–26
novelists with N-strengths, 181–185, 209–210, 229–230, 267–272
Nuance Dragon NaturallySpeaking, 357
nucleus accumbens, 280

O
Oakland Athletics baseball club, 351
O'Brien, Dominic, 215
occupational talents. *See* work and occupational successes

O'Keefe, John, 88
O'Leary, Kevin, 5
oral communication, 264
Orton-Gillingham intervention, 370

P
paper-folding task, 119
paralexic or paraphasic errors (deep substitutions), 149
Party City, 23
pattern completion, 194–195
pattern detection, 138
pattern separation, 193–195
perceiving relationships (similarities and associations), 17, 134–139, 146–150, 166, 170–171
perseverance, 365–366, 371
personal memory. *See* episodic (personal) memory
personal touch skills, 347
perspective shifting, 139–142, 151–152, 155–158, 167
phonemic awareness, 62
phonographic (or pictorial) languages, 71–72
phonological awareness, 62
phonological processing, 47, 54–55, 61–62, 63, 71–72
pi (n), 215
picture production task, 135
place cells, 88–89
Plan B skills, 349
plan creation and review, 341–342
podcasts, as assistive technology, 357
positive attitude, 263, 322, 364–366
prediction
  in action, 270–272, 311
  big-P and small-P predictions, 278–279
  as education principle, 321–323, 326
  error-driven learning and, 279–285, 321–323
  power of, 243–245, 266, 277
  at work and in the workplace, 349

procedural learning and memory
  declarative memory comparison, 57–58, 195–197
  overview of procedural learning, 64–66
  overview of procedural memory, 50–53, 186–187
  trade-offs, 203, 220–222
protein research, 79–80
Pugh, Kenneth, 119–120

Q
qualitative reasoning, 250–251

R
RAND Corporation, 156–157
rapid automatized naming (RAN), 62–63
reading and spelling challenges, 60–72
  brain function and, 36–37, 399n48, 409–410n277
  conclusion, 72
  language processing, 38–40, 62, 71–72, 97–101, 109
  letter reversals, 95–97
  naming speed, 62–63
  new view on, 316–317, 323–327, 370–371, 378, 404–405n155
  overview, 60–62
  phonological processing, 47, 54–55, 61–62, 63, 71–72
  procedural learning and memory, 64–66
  processing speed, 69
  as protective/resilience factors, 69–71
  visual processing, 63–64
  working memory, 66–69
reframing skills, 346–347
relaxation phase, of processing, 253, 255–256, 258
resilience, 365–366, 371
restaging process, 188–190
Reversals (Simpson), 149

Rice, Anne, 181–183, 188–189, 203, 406n181
right hemisphere of brain, 34–40, 398n40, 399n48. *See also* hemispheres of the brain
Rittenberg, Gerry, 22–24, 225–226, 263
Rogers, Richard, 5
Rojkind, Michel, 5
rote memory
   attentional focus and, 283
   metacognition and, 363
   processing of, 50, 65–66, 83
   trade-offs with, 16, 320
Royal Danish Academy of Fine Arts, 111
Russell, James T., 19, 107

S
saccades, 64
SAT (Scholastic Aptitude Test), 122–123, 173, 240, 407n240
"sawtooth" memory profiles, 221
scene-based depictions, 192–195
scene construction, 190–191, 224–225, 250, 255, 277, 319
schemas, 196–197
Schneps, Matthew, 91–92, 117–119, 282
Schoenbrod, David, 92, 200
Schultz, Philip, 226–230
Schwab, Charles, 200–201
Scripps Institution of Oceanography, 308
Sears, 177–178
self-advocacy, 162–163, 347, 374
self-concept. *See* modeling the dyslexic self
self-talk, 333–334
semantic (abstract/impersonal) memory
   key points about, 211–212
   overview, 187–188, 191
   schemas and, 196–197
   trade-offs with, 192–195, 202–203
semantic fields, 38–40, 398n40
Shaywitz, Bennett, 36
Shaywitz, Sally, 36
shifting perspectives, 139–142, 151–152, 155–158, 167
short-term memory, 186
signal-to-noise ratio, 53
similarities and associations (perceiving relationships), 17, 134–139, 146–150, 166, 170–171
Simon, Carly, 5
Simpson, Eileen, 149
simulation for prediction or problem-solving, 244
Slay the Dragon (film), 359
SLI (specific language impairment), 404n144
small-p predictions, 278–279
Smith, Duane, 199–200
Smythe, Ian, 135
social skills, 104
Sowell, Thomas, 402–403n99
spatial imagery, 89
spatial reasoning, 80–81, 82–85, 109, 311, 342–343
spatial-visual creativity, 82–85, 135–136
specific language impairment (SLI), 404n144
spelling. *See* reading and spelling challenges
Spellwright (Charlton), 209, 210
Spencer, Octavia, 5
Sperry, Roger Wolcott, 34
Spielberg, Steven, 5, 142
staff
   for dyslexic employees, 343
   for dyslexic entrepreneurs, 263–264, 311–312
Steffert, Beverley, 86, 135
stimulants, for ADHD, 256–257
storytelling skills. *See* narrative reasoning

strengths
  D-strengths. *See dynamic reasoning*
  I-strengths. *See interconnected reasoning*
  M-strengths. *See material reasoning*
  N-strengths. *See narrative reasoning*
  at work. *See work and occupational successes*
Structured Literacy, 370
submersible deep-sea research vehicle, 309
substitutions (verbal), 148–150
surprise state, 279–281, 284, 318, 321
survey. *See MIND strengths survey*
Sydney Opera House, 111–112
syllogisms, 99–100
symbols, struggles with, 94–97, 109
Symmes, Jean, 85
systems reasoning, 344

T
Taylor, Helen, 298–306
teamwork facilitation, 344–345, 408n264
tenacity, 309, 337, 365–366
Term Limits (Flynn), 269
text-to-speech devices, 233
Thinking Like Einstein (West), 86–87
A Thousand Brains (Hawkins), 43, 279
3-D spatial reasoning, 80–81, 88–89
Through the Looking-Glass (Carroll), 189
Titanic (passenger ship), 306, 311
togetherness (association), 135, 138–139, 147–150
Tolkien, J. R. R., 380
top-down learners, 152–153
transparency, 72
triage skills, 348–349
Trinity (Uris), 269
true narratives, 245
20,000 Leagues Under the Sea (film), 308

U
Ullman, Michael, 220
ultrasounds, 123–124
Unlimited Memory (Horsley), 215
Uris, Leon, 269
Use Your Memory, 213–214
Utzohn, Jørn, 110–112

V
Virgin Atlantic, 264
visual attention, 63, 282, 305
visual processing
  brain function and, 35
  reading challenges and, 63–64
  visual-spatial creativity, 82–85, 135–136
visual span, 63–64

W
West, Thomas G., 36, 86–87
The Wherewithal (Schultz), 229–230
Wolf, Maryanne, 36, 87, 326, 398n36
Wolff, Ulrika, 86
word sound (phonological) processing, 54–55, 61–62, 63
work and occupational successes, 339–360
  about, xviii–xix, 5, 339–340, 341, 401n86
  additional support for, 263–264, 311–312, 343, 355–360
  conclusion, 351–352, 379
  fields for D-strengths, 341, 348–350
  fields for I-strengths, 341, 343–346
  fields for M-strengths, 340–343, 402–403n99
  fields for N-strengths, 346–348

work and occupational
    successes (*cont.*)
  people with D-strengths in action,
    89, 184, 239–243, 246–250,
    259–261, 267–272, 285–293
  people with I-strengths in action,
    127–132, 138, 154–158,
    160–163, 172–178, 200–201
  people with M-strengths in action,
    75–81, 84–85, 105–106,
    110–112, 121–124, 341–343
  people with N-strengths in action,
    181–183, 207–210, 226–235

working memory
  development and, 99–100,
    104–106, 169–170, 365
  distractions and, 150
  overview, 51–52, 66–69
  trade-offs with, 304
working memory overload, 52
World Memory Championships,
  214–215

Y
Yohana (company), 291
"young engineers," 11–12

About the Authors

Drs. Brock and **Fernette Eide** are leading experts in the field of dyslexia and cofounders of the nonprofit Dyslexic Advantage and the social-purpose corporation Neurolearning.com. They have served as consultants to the President's Council on Bioethics and as visiting lecturers at the Stanford Graduate School of Education. The first edition of their book, *The Dyslexic Advantage*, was an international bestseller.